How Everything Came About

How Everything Came About

from Big Bang to Earth, Life and Technology

Peter Kenny

The most incomprehensible thing about the world
is that it is comprehensible.
The fact that it is comprehensible is a miracle.
                                                    Albert Einstein

Copyright © Peter Kenny, 2017

Available from Amazon.com, CreateSpace.com and other retail outlets

Available on Kindle and other devices

Printed by CreateSpace

Cover Design by Louise Flanagan

ISBN 9781541387966

Dedicated to the memory of my parents, Doris and William Kenny, who taught me the value of education, having had little of it themselves.

By the Same Author:

A Handbook of Public Speaking for Scientists and Engineers

Decisions from Data

Better Business Decisions from Data

Preface

To explain in detail how everything came about from the Big Bang to the present day would require more than a library of books, and most of us would find much of the information difficult if not impossible to follow. We each have our own areas of specialised knowledge and beyond that we are content to have a basic understanding of a wider range of subjects.

This book provides, in concise units of information, the briefest account of how everything we see around us came into existence. It is a story worth telling and appreciating because it is so amazing. The Big Bang, fourteen billion years ago, produced a few tiny particles. These responded to the influence of just four forces to produce atoms, molecules, and materials that constitute everything we encounter. Furthermore, the same building blocks and forces produced life, humans, awareness, intelligence and accelerating technological progress.

It is not a book to be read from beginning to end, but rather a book to dip into. The way I have presented the contents, as a route map, makes this clear. From a planned or casual entry, anywhere, it is easy to progress forwards or backwards in order to trace a theme. Readers unfamiliar with a topic may find the information adequate. Others may be encouraged to explore a topic further. Some may find reminders of facts once known but now vague in the memory. Students may find useful revision notes within the pages.

In defining and choosing the topics and the information they contain, I have tried to avoid dealing with matters that are obvious, or generally considered to be obvious; even if in reality there are commonly unperceived subtleties (as there often are). Some topics that would be considered difficult, such as quantum mechanics and relativity, have been included because they are fundamental in the way the Universe operates. Also, there is much curiosity surrounding such subjects, and considerable misunderstanding. To illustrate technological progress I have concentrated on topics such as lasers and nanotechnology because of their importance for the future and the prominence that is given to them in the media.

I have been generous with index entries on the grounds that it is useful to appreciate the context of an item even though the information given may be minimal.

If you find yourself reminded of attempts to present a Shakespeare play in three minutes flat, I am not too surprised. However, whereas Romeo and Juliet in three minutes would lose all that is great about Shakespeare and merely preserve a trivial plot, I believe that here you will find we have lost detail of a more specialist nature while preserving the much more interesting essentials.

Finally, if you find any inaccuracies, or indeed if you have any comments or suggestions, I would be very pleased to hear from you.

Peter Kenny
Lichfield,
January 2017
kenny.peter@physics.org

How Everything Came About

Contents and Route Map

How Everything Came About

1. The First 400,000 Years
2. Expansion of the Universe
3. Galaxies
4. The Milky Way
5. Stars
6. The Solar System
7. The Moon
8. The Earth
9. Landforms
10. Rocks
11. The Atmosphere
12. Climate
13. Weather
14. Earth History
15. Evolution
16. Evolution of Humans
18. Photons
19. Light
20. Lasers
30. Radioactivity
31. Nuclear Energy
44. Chemical Processes
45. Life
46. Origin of Life
47. Extraterrestrial Life
48. Bacteria and Viruses
49. Plants
50. Animals
51. The Human Frame
52. Body Circulation System
53. The Immune System
54. Food
55. Human Metabolism
56. Reproduction
57 Genetics
58. The Endocrine System
59. The Nervous System
60. The Senses
61. The Brain

How Everything Came About

17. Elementary Particles

29. Atomic Nuclei

21. Electrons

32. Atoms

22. Electric Currents

33. Molecules

23. Magnetism

36. Liquids

24. Electric Power

37. Liquid Crystals

25. Electronic Systems

38. Glasses

26. Electronic Components

39. Polymers

40. Metals and Alloys

27. Electronic Processing

41. Crystalline Solids

28. Semiconductors

42. Composite Materials

43. Nanotechnology

34. Sound

35. Heat and Temperature

67. General Relativity

68. Special Relativity

62. Quantum Mechanics

69. The Twin Paradox

63. String Theory

64. Quantum Entanglement

70. Why is the Universe as it is?

65. Quantum Computing

71. Is the Future Determined?

66. Quantum Cryptography

I. The First 400,000 Years

The Universe came into existence about fourteen billion years ago. A very small and very hot speck of energy expanded in the form of radiation. As it expanded it cooled, and some of the energy was able to convert to mass, appearing as small particles. With further expansion and cooling the particles were able to combine.

When the Universe was about three minutes old hydrogen and helium nuclei were present together with electrons, photons and neutrinos. After about 400,000 years, the Universe had cooled sufficiently to allow electrons to combine with the nuclei to form atoms, without being dislodged by energetic photons. In effect the Universe became transparent as the photons were now able to move freely through space. Photons from this period constitute the cosmic background radiation which can still be detected over the entire sky.

What was there before the Big Bang? Why did it happen? We have no way of answering these questions. We do know that time and space as we experience them were products of the Big Bang so there was no 'before' or 'elsewhere' in the way we would understand the terms.

There is, of course, speculation involving for example, previous universes or parallel universes, which we would never be able to observe.

2. Expansion of the Universe   17. Elementary Particles
62. Quantum Mechanics   67. General Relativity
                70. Why is the Universe as it is?

How Everything Came About

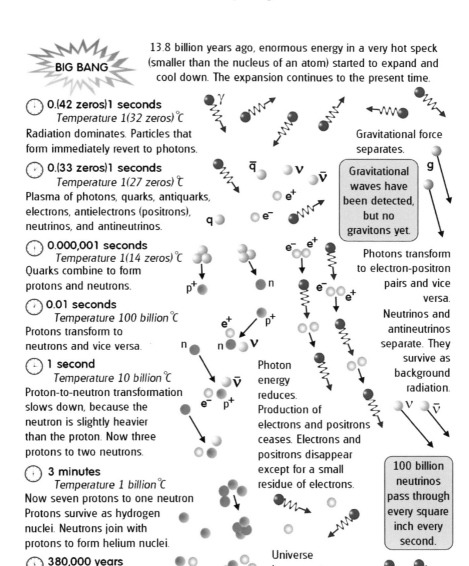

BIG BANG — 13.8 billion years ago, enormous energy in a very hot speck (smaller than the nucleus of an atom) started to expand and cool down. The expansion continues to the present time.

0.(42 zeros)1 seconds
Temperature 1(32 zeros)°C
Radiation dominates. Particles that form immediately revert to photons.

0.(33 zeros)1 seconds
Temperature 1(27 zeros)°C
Plasma of photons, quarks, antiquarks, electrons, antielectrons (positrons), neutrinos, and antineutrinos.

0.000,001 seconds
Temperature 1(14 zeros)°C
Quarks combine to form protons and neutrons.

0.01 seconds
Temperature 100 billion °C
Protons transform to neutrons and vice versa.

1 second
Temperature 10 billion °C
Proton-to-neutron transformation slows down, because the neutron is slightly heavier than the proton. Now three protons to two neutrons.

3 minutes
Temperature 1 billion °C
Now seven protons to one neutron. Protons survive as hydrogen nuclei. Neutrons join with protons to form helium nuclei.

380,000 years
Temperature 3000°C
Nuclei combine with electrons to form hydrogen and helium atoms.

(The proportion of hydrogen to helium (3 to 1 by weight) is retained throughout the universe at the present time.)

From this time the universe is dominated by gravity and matter, and not radiation.

Gravitational force separates.

Gravitational waves have been detected, but no gravitons yet.

Photons transform to electron-positron pairs and vice versa.

Neutrinos and antineutrinos separate. They survive as background radiation.

Photon energy reduces. Production of electrons and positrons ceases. Electrons and positrons disappear except for a small residue of electrons.

100 billion neutrinos pass through every square inch every second.

Universe becomes transparent. Photons separate.

The photons survive as the cosmic background radiation. The wavelength has stretched (cooled) and is now in the microwave region, equivalent to 2.73°C above absolute zero.

How Everything Came About

1. The First 4000,000 Years

2. Expansion of the Universe

The expansion of the Universe has continued since the Big Bang to the present time. Space is stretching in all directions so that galaxies that are farther away are moving away from us at a greater speed. There is no centre of the Universe: the Big Bang was everywhere.

There are three possibilities for the future. Gravity may slow down the expansion sufficiently to cause a contraction, ending in the 'Big Crunch'. At the other extreme, the expansion may continue forever, resulting in an ever-expanding cold, dead Universe. The in-between option is expansion at an ever-decreasing rate so that a limiting size is gradually approached.

Evidence points to the latter scenario, though the expansion is currently speeding up rather than slowing down. An increasing rate of expansion implies an antigravity effect.

It appears that about two thirds of the total energy in the Universe is 'dark energy', which repels matter. Of the remaining third, most is 'dark matter' which is subject to gravitational attraction, but the constitution of which is not known for sure. The ordinary material that we observe makes up a minor part only.

3. Galaxies

How Everything Came About

Expansion of space started at the Big Bang and still continues. The expansion is of space and not into space, i.e. there is nothing outside. Because space is stretching and carrying galaxies with it, the speed of expansion is not restricted to the speed of light.

Big Bang plus 0.(42 zeros)1 seconds
Size less than one atom.

0.(34 zeros)1 seconds (Inflation Period)
Very rapid expansion. The size of a melon now encompasses all of the Universe we can ever observe.

Gravity slows down the expansion.

7 billion years
The rate of expansion starts to increase. This antigravity effect is attributed to dark energy.

Only 5% of the matter in the Universe is ordinary. 27% is dark matter which is subject to gravitational attraction. 68% is dark energy which repels matter and is causing the expansion to speed up. It is not know what dark matter or dark energy are.

Note that space is stretching in all directions, so there is no central point. Any galaxy cluster may appear to be at the centre, since all the others are receding from it.

Now - 13.8 billion years
The rate of expansion is still increasing. Clusters of galaxies are moving away from each other. Those twice as far away are moving twice as fast, and so on (Hubble's Law). The size of the Universe is not known and may be infinite.

The Universe has a flat geometry (straight lines are truly straight). This implies that the total mass is close to critical, eventually giving ever-slowing expansion. With greater than critical mass, gravity would reverse the expansion and collapse the Universe in a final 'Big Crunch'. With less than critical mass, the Universe would grow without limit.

2. Expansion of the Universe

3. Galaxies

As the Universe expanded, the hydrogen and helium were pulled by gravity into clouds which compacted sufficiently to form stars. Gravitational attraction between stars held groups of them together in galaxies. The galaxies themselves formed clusters which are held by gravity and do not expand with the general expansion of the Universe. It is the superclusters that move apart.

Galaxies are constantly changing as new stars are formed and old ones die. Merging of galaxies occurs when gravity draws them together. The stars in galaxies rotate about the galaxy centre where there is considered to be a massive black hole.

We are now able to see galaxies more than 13 billion light years away. We are seeing them as they were 13 billion years ago because of the time taken for the light to reach us. This is close to the limit of the observable Universe because the Big Bang was 13.8 billion years ago. The present size of the Universe is not known.

Our own galaxy, the Milky Way, is the second largest of more than 50 galaxies in our Local Group, Andromeda being the largest.

The galaxies in the observable Universe, 2 trillion or so, are more densely packed than the stars in the galaxies.

4. The Milky Way

How Everything Came About

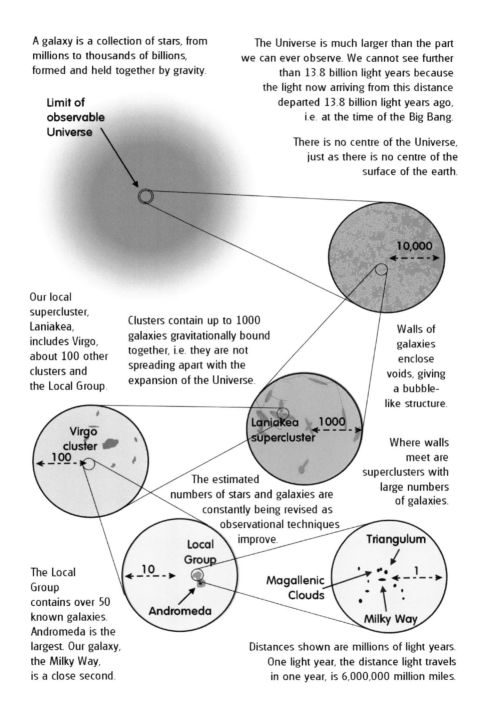

How Everything Came About

3. Galaxies

4. The Milky Way

Our galaxy, the Milky Way, is a spiral galaxy, disc-like in shape with a central bulge. It is about 100,000 light years across and holds about 300 billion stars.

    The stars of the galaxy are rotating around the centre where there is a black hole. Stars closer to the centre rotate faster. Our solar system is moving at 560,000 miles per hour and takes 230 million years to go once around.

    Our nearest neighbouring star is four light years away. This distance is about 5000 times the size of our solar system.

5. Stars

How Everything Came About

The Milky Way, our galaxy, is a spiral galaxy, disc-like in shape with a central bulge. It holds about 300 billion stars and has a black hole at the centre as do other galaxies.

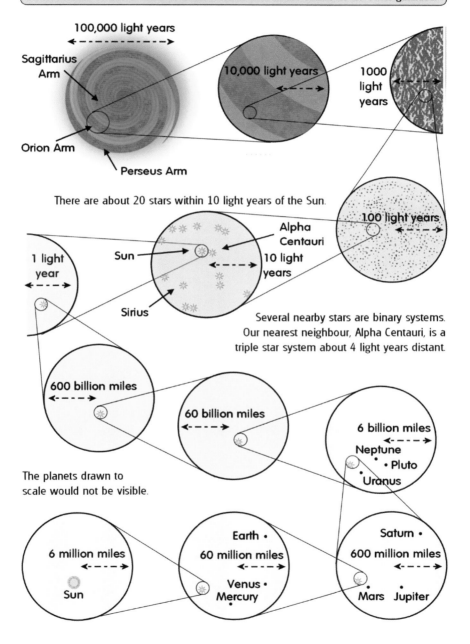

There are about 20 stars within 10 light years of the Sun.

Several nearby stars are binary systems. Our nearest neighbour, Alpha Centauri, is a triple star system about 4 light years distant.

The planets drawn to scale would not be visible.

4. The Milky Way

5. Stars

Stars are formed as clouds of gas contract under gravity and heat up because of the compression.

The life history and lifetime of a star depends on the amount of material it holds. Small stars simply contract to solid brown dwarfs. Larger ones become hot enough to start fusion of hydrogen to form helium, which releases energy in the form of radiation.

This stage, the main sequence, occupies most of the lifetime of a star, the bigger the star the shorter the time. When all the hydrogen has been used up, the star expands to a red giant and if the star is large enough helium fusion starts.

In the largest stars helium fusion is followed by fusion of progressively heavier elements up to iron, at which point fusion stops. The star cools and collapses and then explodes in a supernova. In the explosion elements heavier than iron are formed and scattered into space. This material adds to the gas used in the formation of the next generation of stars and accounts for the presence of the range of elements found on Earth.

Final collapse may form a black hole. Here gravity is so strong because of the large mass of material that nothing, not even light, can escape.

6. The Solar System

How Everything Came About

(Not to scale)   The first stars formed about 200 million years after the Big Bang.

CLOUD OF GAS   Clumps of gas, mainly hydrogen, are drawn together by gravity and heat up as they compress.

Larger clumps reach a temperature of 15,000,000° C. Fusion of hydrogen to helium starts.

Massive   Small
Hot   Cool   **BROWN DWARF**

MAIN SEQUENCE STARS

Stars spend most of their lives as main sequence stars, ranging from a few million years for the largest stars to billions of years for the smallest. As hydrogen fusion ceases and helium fusion begins, the star expands to a red giant.

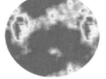

RED GIANTS

When hydrogen fusion stops, outer layers of gas are shed.

Fusion produces progressively heavier elements, but cannot continue beyond iron.

If the mass is more than 3 times our Sun's mass, final collapse results in a black hole.

PLANETARY NEBULA

BLACK HOLE

Gravity is so intense that spacetime curves back on itself. Nothing, not even light, can escape from a black hole.

If the mass is more than 1.4 times our Sun's mass, final collapse results in a neutron star.

WHITE DWARF

SUPERNOVA

NEUTRON STAR

Intense gravity forces electrons to combine with protons, producing tightly packed neutrons.

BLACK DWARF

Final collapse causes an explosion which provides the energy to form, by fusion, all elements heavier than iron and to scatter them through space. A neutron star or black hole remains.

5. Stars

6. The Solar System

Our Sun is a fairly small star located towards the edge of our galaxy. It is about 4600 million years old, and about half way through is lifetime as a main sequence star. About 99.9% of the mass of the solar system is in the Sun.

Nine planets rotate about the Sun in approximately the same plane, hence they are seen to follow the path of the Sun in the sky. It was a matter of some dispute whether Pluto, so small and far out, should continue to be considered a planet. Its orbit is very elongated. It is now classed as a dwarf planet.

The four inner planets are solid though Venus is covered in dense clouds so the surface cannot be seen directly. The four giant planets consist mainly of gas.

The planets rotate in approximately the same plane as they rotate round the Sun, with the exception of Uranus which rotates at right angles to its orbital rotation.

Asteroids are smalt rocky pieces in orbit while comets consist of ice and dust. The characteristic comet's tail is material being evaporated by the heat of the Sun. The orbits of comets are extremely elongated so they spend most of their time a long way from the Sun.

7. The Moon

How Everything Came About

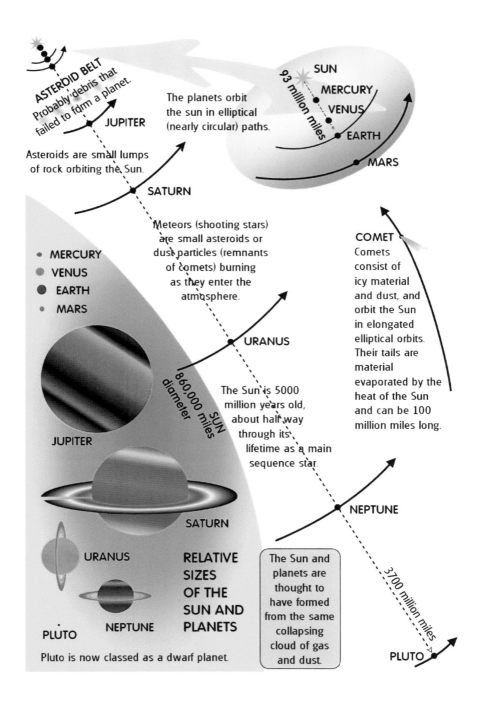

ASTEROID BELT Probably debris that failed to form a planet.

Asteroids are small lumps of rock orbiting the Sun.

The planets orbit the sun in elliptical (nearly circular) paths.

93 million miles

SUN, MERCURY, VENUS, EARTH, MARS

JUPITER

SATURN

Meteors (shooting stars) are small asteroids or dust particles (remnants of comets) burning as they enter the atmosphere.

- MERCURY
- VENUS
- EARTH
- MARS

URANUS

COMET Comets consist of icy material and dust, and orbit the Sun in elongated elliptical orbits. Their tails are material evaporated by the heat of the Sun and can be 100 million miles long.

860,000 miles SUN diameter

The Sun is 5000 million years old, about half way through its lifetime as a main sequence star.

JUPITER

SATURN

NEPTUNE

URANUS

RELATIVE SIZES OF THE SUN AND PLANETS

The Sun and planets are thought to have formed from the same collapsing cloud of gas and dust.

PLUTO   NEPTUNE

Pluto is now classed as a dwarf planet.

3700 million miles

PLUTO

6. The Solar System

7. The Moon

Earth's satellite, the Moon, is one of many (about 200) satellites present in the solar system but it is somewhat unusual in being an appreciable size compared to the Earth. This is because it was not formed from asteroids and debris captured by the Earth, but as a result of an asteroid collision ejecting material from the Earth. Rather remarkably, its size and distance from the Earth result in it appearing to be almost exactly the same size as the Sun in the sky.

It is illuminated by the light from the Sun and its position relative to the Sun produces the visual effects of phases and eclipses. Its position and its gravitational attraction gives rise to the ocean tides on Earth.

The same side of the Moon always faces the Earth. This would not have been so originally, but the mutual tidal pull between the Earth and Moon has slowed the Moon's rotation and locked it in position.

The low mass of the Moon gives insufficient gravity to retain an atmosphere. Its surface is unweathered and shows the scars of intensive asteroid bombardment experienced in the period following its formation.

Volcanic activity, present in the past, has now ceased.

8. The Earth

How Everything Came About

Soon after the Earth formed, an asteroid collided, causing ejection of material which solidified as the Moon. Its diameter, 2160 miles, is relatively large compared to the Earth.

(Not to scale)

The Moon orbits Earth once a month, our view of the illuminated side appearing as the phases of the Moon.

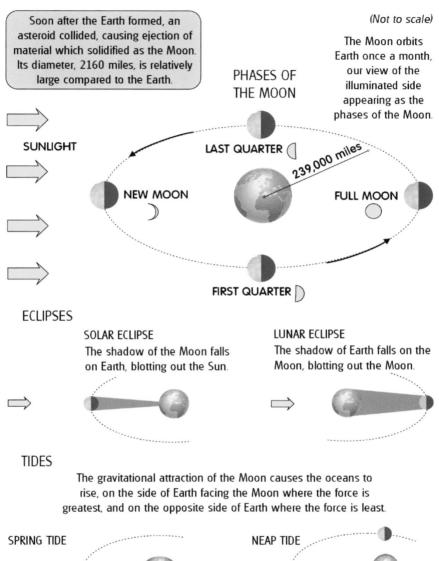

PHASES OF THE MOON

SUNLIGHT

LAST QUARTER

NEW MOON

239,000 miles

FULL MOON

FIRST QUARTER

ECLIPSES

SOLAR ECLIPSE
The shadow of the Moon falls on Earth, blotting out the Sun.

LUNAR ECLIPSE
The shadow of Earth falls on the Moon, blotting out the Moon.

TIDES

The gravitational attraction of the Moon causes the oceans to rise, on the side of Earth facing the Moon where the force is greatest, and on the opposite side of Earth where the force is least.

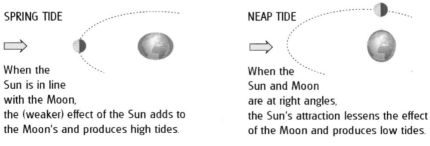

SPRING TIDE

When the Sun is in line with the Moon, the (weaker) effect of the Sun adds to the Moon's and produces high tides.

NEAP TIDE

When the Sun and Moon are at right angles, the Sun's attraction lessens the effect of the Moon and produces low tides.

How Everything Came About

7. The Moon

8. The Earth

As gravity pulled material together to form the Earth, heavier constituents tended to migrate to the centre. The centre is at a very high temperature but the weight of overlying rock renders the material solid. Outside this solid inner core, where the pressure is less, the iron-rich material is liquid. Convection currents in the liquid are responsible for generating the Earth's magnetic field.

Heat transfer to the surface of the earth creates convection currents below the surface which cause the surface rocks to move and fracture.

The surface of the Earth has split into a number of large plates and it is these that move relative to each other under the influence of the convection currents. As plates are pushed together and pulled apart, mountains and mid-ocean ridges are formed, earthquakes occur and volcanoes appear.

The continents have thus occupied different positions on the Earth's surface, once coming together as one large continent.

The surface rock masses are lighter than the underlying rocks and in effect are floating on them, like icebergs floating on the sea. As surface rock is weathered away and washed into the oceans, the land mass uplifts.

9. Landforms   11. The Atmosphere   14. Earth History

How Everything Came About

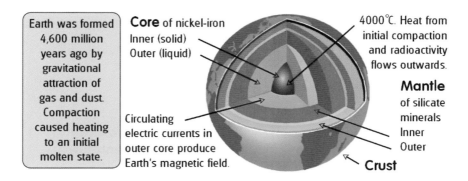

Earth was formed 4,600 million years ago by gravitational attraction of gas and dust. Compaction caused heating to an initial molten state.

Core of nickel-iron
Inner (solid)
Outer (liquid)

Circulating electric currents in outer core produce Earth's magnetic field.

4000°C. Heat from initial compaction and radioactivity flows outwards.

Mantle of silicate minerals
Inner
Outer

Crust

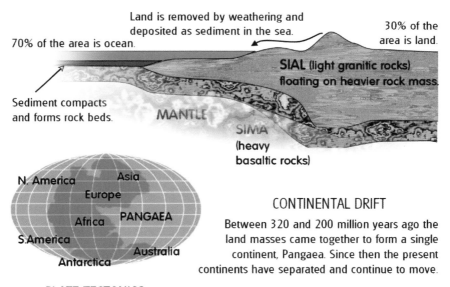

Land is removed by weathering and deposited as sediment in the sea.

70% of the area is ocean.

30% of the area is land.

SIAL (light granitic rocks) floating on heavier rock mass.

Sediment compacts and forms rock beds.

MANTLE

SIMA (heavy basaltic rocks)

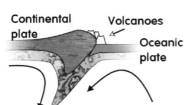

N. America, Asia, Europe, Africa, PANGAEA, S.America, Australia, Antarctica

CONTINENTAL DRIFT

Between 320 and 200 million years ago the land masses came together to form a single continent, Pangaea. Since then the present continents have separated and continue to move.

PLATE TECTONICS

Earth's surface is made up of plates of crust and upper mantle material. Convection currents cause the plates to move relative to each other.

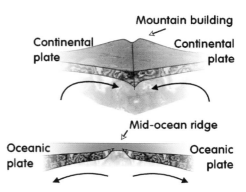

Continental plate
Volcanoes
Oceanic plate

Mountain building
Continental plate — Continental plate

Mid-ocean ridge
Oceanic plate — Oceanic plate

8. The Earth

9. Landforms

The presence of an atmosphere on Earth gives rise to weather and the weathering processes gradually remove the land surface, depositing the material in the oceans. By now, the Earth would be covered with a layer of water were it not for the land-uplifting processes.

The breaking down of the land mass into small pieces occurs by chemical and physical processes. The movement of the material to the sea is driven by gravity the transport being by river, glacier or wind

The sea itself attacks coastal regions, powered by wind and tide. The effect is enhanced by the changes in sea level as regions of the land mass rise or sink.

Uplift of land is brought about in mountain building resulting from land masses moving together. In addition, a buoyancy effect causes the land to rise as its weight is reduced by the loss of material to the oceans. Molten rock, brought to the surface by volcanic activity, adds to the land mass.

10. Rocks

How Everything Came About

Landforms result from internal (endogenic) processes, which increase the land elevation, and external (exogenic) processes which, on average, decrease elevation and relief.

ENDOGENIC PROCESSES

Volcanic activity brings molten rock to the surface.

Converging tectonic plates create mountains.

As material is removed from continents, the land rises because of the buoyancy effect (isostasy).

EXOGENIC PROCESSES

CHEMICAL WEATHERING

Rock minerals are altered and removed by environmental attacks, such as solution, oxidation or carbonation.

Gravity and weathering act in conjunction with other exogenic processes, but sometimes act alone, as in the formation of debris slopes (screes) at the bases of hills.

PHYSICAL WEATHERING

Volume changes due to thermal cycling, or to ice or salt crystals opening cracks, fractures rock surfaces.

RIVERS

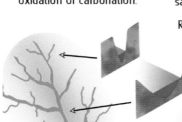

Rivers tend to form a dendritic pattern but the underlying rock strata may create parallel streams. Previous topography also has an effect. Erosion at the bottom of the stream, where the debris is carried, forms a V-shaped section.

Delta

Deposition at the mouth of a river may form a delta.

GLACIERS

Glaciers are slow moving rivers of ice. Abrasion by debris within the ice, and plucking of material from the sides and bottom of the valley, create a U-section. The glacier terminates in the sea, breaking up as icebergs, or on land, depositing mixed size debris.

WAVES

Erosion by waves creates cliffs and shore platforms. Deposition by waves creates beaches, spits and barrier islands. Many coastal features are the result of changes in sea level.

Large areas of the earth have been glaciated in the past (ice ages) and exhibit post-glacial features from the scouring of the land and deposits of material.

WIND

Erosion by wind-borne debris, producing vertical rock faces, pebbles and sand, is important where there is lack of vegetation. The finest material is blown away (deflation). Sand dunes form in deserts.

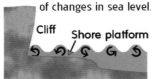

Cliff   Shore platform

9. Landforms

10. Rocks

As the surface of the Earth cooled from its original molten state, elements combined to form minerals, mainly oxides, and the minerals made up the solid rock.

Igneous rocks, such as granite, are material that has cooled and solidified in this way. They consist of minerals that have grown as crystals and locked together in the solid rock mass. The slower the cooling the larger the individual crystals. Originally all rock on the surface of the earth would have been igneous.

Weathering processes gradually wear down the land surface and deposit the debris in the sea, in layers. These layers become compacted and eventually uplifted, forming beds of sedimentary rocks such as sandstone. Sedimentary rocks vary in grain size which reflects the nature of the original debris.

Sedimentary rocks may come into contact with molten rock and be altered by the heat and by pressure. These altered rocks are the metamorphic rocks such as slate, which is metamorphosed shale.

Some rocks are organic in origin. Coal is vegetation that has been subject to heat and pressure. Some limestones are the remains of shells of sea creatures.

Although rocks can be conveniently grouped into characteristic types, there is a continuous spectrum of variation of composition and crystal or grain size.

How Everything Came About

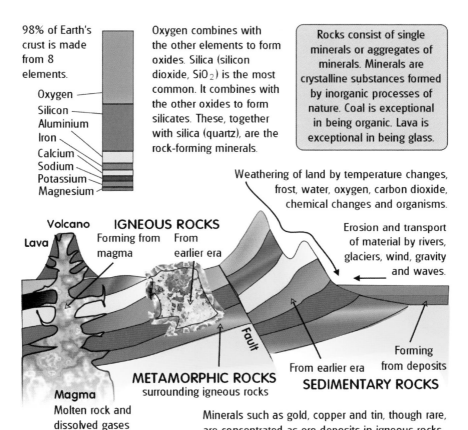

98% of Earth's crust is made from 8 elements.
- Oxygen
- Silicon
- Aluminium
- Iron
- Calcium
- Sodium
- Potassium
- Magnesium

Oxygen combines with the other elements to form oxides. Silica (silicon dioxide, SiO_2) is the most common. It combines with the other oxides to form silicates. These, together with silica (quartz), are the rock-forming minerals.

Rocks consist of single minerals or aggregates of minerals. Minerals are crystalline substances formed by inorganic processes of nature. Coal is exceptional in being organic. Lava is exceptional in being glass.

Weathering of land by temperature changes, frost, water, oxygen, carbon dioxide, chemical changes and organisms.

Erosion and transport of material by rivers, glaciers, wind, gravity and waves.

IGNEOUS ROCKS — Forming from magma / From earlier era

Volcano, Lava

METAMORPHIC ROCKS surrounding igneous rocks

SEDIMENTARY ROCKS — From earlier era / Forming from deposits

Magma — Molten rock and dissolved gases

Fault

Minerals such as gold, copper and tin, though rare, are concentrated as ore deposits in igneous rocks.

IGNEOUS ROCKS Solidified molten rock	**BASALT** Brought to the surface by volcanic activity, rapid cooling giving fine crystal structure or glass (lava).		**GRANITE** Slowly cooled at depth giving a structure of large crystals.
METAMORPHIC ROCKS Pre-existing rock affected by heat and/or pressure	QUARTZITE ↑	SLATE ↑	MARBLE ↑
SEDIMENTARY ROCKS Compressed debris from weathering of pre-existing rock	SANDSTONE Coarse grained	SHALE Fine grained	**LIMESTONE** Shells of sea creatures and precipitates from water.
	PEAT, COAL Remains of vegetation		

How Everything Came About

8. The Earth

11. The Atmosphere

The Earth is sufficiently large for its gravitational force to retain an atmosphere. It consists mainly of nitrogen and oxygen with small amounts of carbon dioxide and other gasses. It is most dense close to the surface of the Earth and most of it lies within a few miles of the surface, within the troposphere.

Radiation from the Sun, consisting of photons and ionised gas, interacts with the upper atmosphere and produces numerous effects. The region of ions and free electrons, the ionosphere, affects the propagation of radio waves. The Earth's magnetic field funnels charged particles to the polar regions and produces the aurora.

Ultraviolet photons interact with oxygen and produce ozone, which, in turn, protects the Earth from the biologically damaging ultraviolet radiation.

The greenhouse effect results in a warming of the surface of the Earth. It arises because incident infrared radiation from the sun penetrates the atmosphere, while the radiation emitted by the Earth back into space is of lower energy and is absorbed by the atmosphere. The atmosphere protects the Earth from meteors, causing all but the largest to burn away by friction as they enter.

12. Climate

How Everything Came About

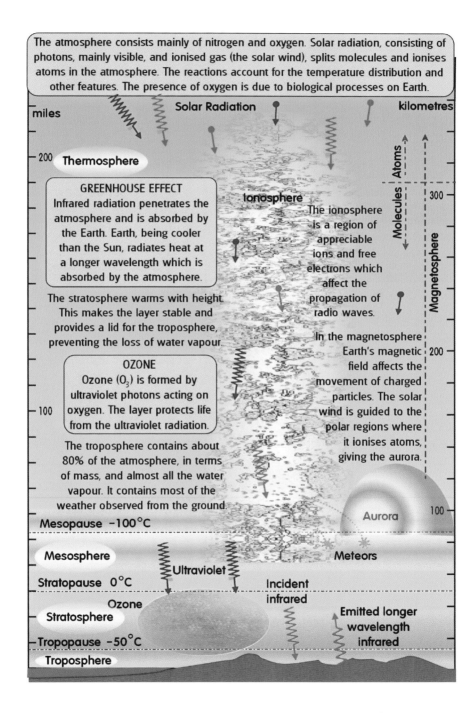

11. The Atmosphere

12. Climate

The Earth's rotation about its axis is nearly in the same plane as its rotation around the Sun, so the equatorial regions constantly receive more radiated heat from the Sun than do the polar regions. This sets up convection currents in the lower atmosphere. The resulting winds transport warm and cool air across the surface.

The winds are deflected to the east or west as they travel north or south, because of the Coriolis Effect. The equatorial regions are moving eastwards at about 1000 miles per hour as the Earth rotates. Regions north or south are moving more slowly, so north-south winds have an east-west speed differing from that of the land surface below and are therefore deflected. The direction of deflection is opposite in the two hemispheres.

Because the Earth's rotation is not exactly in line with its rotation around the Sun, regions north and south of the equator receive alternating higher and lower heating from the Sun in the course of a year. This produces seasonal variations in the climate.

The presence of the oceans influences climate because the water warms and cools more slowly than the land, the radiation from the Sun penetrating to a greater depth in water than in the land cover.

13. Weather

How Everything Came About

Climate is the long-term average weather and its variability. It affects vegetation and is often classified in terms of the vegetation rather than the meteorological factors.

THE SEASONS

The heat of the Sun acting on the land, sea and atmosphere produces weather. The heating effect is greater in the summer months because of the tilt of Earth's axis of rotation.

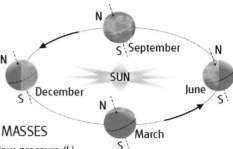

PREVAILING MOVEMENT OF AIR MASSES

Rising warm air at the equator creates low pressure (L). Descending cold air at the poles creates high pressure (H). Air moves from high to low pressure regions but is deflected sideways by the Coriolis effect. The net result is several major convection cells. The Ferrel cells, shown idealised, are not directly thermally driven. The north cell is variable and affected by cyclones and anticyclones.

CORIOLIS EFFECT
Air moving north or south is deflected sideways because equatorial regions are rotating faster towards the east than are the higher latitudes.

The south one is more constant, the scarcity of land mass inhibiting stationary pressure systems.

WORLD CLIMATIC REGIONS

Various classification systems have been employed, commonly based on temperature and aridity (precipitation less evaporation). The diagram shows a simplified broad outline.

- ■ **Polar** Winters long, cold to severely cold.
- ■ **Cold** Winters snowy, cold to very cold. Summers warm to very cold.
- □ **Temperate** Rainy, seasonal. Winters cool.
- ■ **Desert and Steppe** Hot and dry but including continental interiors having cool winters.
- ■ **Tropical** Hot, wet. Little seasonal change but some regions have wet and dry seasons.
- ■ **Highland** Variable, generally cooler and wetter than nearby lowlands.

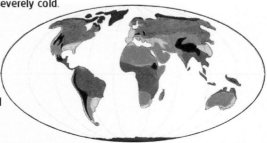

12. Climate

13. Weather

Within any climatic region there are short-term and local variations in the state of the atmosphere.

Warm air passing over the sea absorbs moisture. If the air rises and cools the moisture is released as clouds, rain, hail or snow.

Local cool regions cause air to descend giving high pressure and outflowing winds. These, deflected by the Coriolis Effect, produce rotating wind systems called anticyclones which tend to be stable and slow moving. Clear sky and low wind speed are characteristic.

Depressions arise where low pressure warm air meets high pressure cool air. Warm air becomes trapped in a mass of cool air and the resulting low pressure system rotates in the opposite direction to that of an anticyclone. Warm and cold fronts with accompanying precipitation separate the warm and cool air.

Electrical effects arise in the atmosphere when atoms become ionised by strong air movements, the negatively charged electrons separating from the positively charged ions. Lightning is the consequent electrical discharge.

Wind speed can be expressed by a number on the Beaufort scale, ranging from 1 for calm conditions up to 12 for a hurricane.

How Everything Came About

Weather is the local state of the atmosphere in terms of temperature, wind, humidity, cloud, precipitation and air pressure. While latitude and season determine the broad features, local air movements account for the detailed effects.

Air expands and cools on rising. Water vapour condenses to give cloud, rain, hail or snow.

Warm air absorbs water vapour from the sea.

Cool air forces warm air to rise above.

Mountains force air to rise.

Surface heating warms the air which then rises

The sun's rays penetrate the sea to greater depth than the land. The sea warms and cools slowly because of the effects of deep penetration and mixing. Ocean currents, driven by the wind, affect the sea temperature.

The land warms and cools quickly giving extremes of temperature away from coastal influences.

ANTICYCLONE

High pressure from descending cold air causes outward air flow which, with Coriolis deflection, results in rotation (clockwise in the northern hemisphere). Anticyclones are stable and slow moving, giving low wind speed, dry weather, clear sky and fog in winter.

Isobar

DEPRESSION

At a meeting of cold and warm air (a front) a wave of cold air enters the warm air, trapping a mass of low pressure warm air between a cold front and a warm front. The Coriolis effect gives an anticlockwise (northern hemisphere) rotation. Precipitation accompanies the warm and cold fronts.

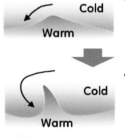

Isobars are lines joining points of equal pressure and appear as contours on weather maps.

THUNDERSTORMS

Strong updraughts and associated downdraughts separate electrical charges (electrons and ions) on particles of water and ice. Lightning is the resulting electrical discharge, usually within clouds but sometimes to earth. Tornadoes develop on the periphery of severely rotating thunderstorms.

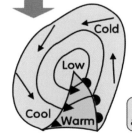

Cyclones, tropical storms, typhoons and hurricanes are severe depressions.

BEAUFORT SCALE

	miles/hour
1 Calm	
2 Light air	
3 Gentle breeze	12
4 Moderate breeze	
5 Fresh breeze	24
6 Strong breeze	
7 Near Gale	
8 Gale	46
9 Strong gale	
10 Storm	63
11 Violent storm	72
12 Hurricane	

How Everything Came About

8. The Earth

14. Earth History

About 4600 million years ago the Earth formed from rocks and frozen gases pulled together and heated by gravitational attraction.

Once the surface had solidified and the atmosphere and oceans had been established, weathering began to lay down sedimentary rocks. These contain fossil remains of life forms. Fossils older than 570 million years ago, of the Precambrian era, are microscopic and were not recognised as fossils until relatively recently.

The early fossils indicate that life started relatively soon after the Earth formed, prior to oxygen being released into the atmosphere from the sea. Life forms were single-cell creatures and it was a further 3000 million years before multi-cellular life developed.

The Cambrian period saw the sudden appearance of numerous sizeable creatures having shells. The reason for this 'Cambrian Explosion' is not known though there are many hypotheses. Increases in oxygen level or nutrients, temperature change, or genetic changes giving the ability to produce hard body parts or to produce a larger body size, could be responsible.

15. Evolution

How Everything Came About

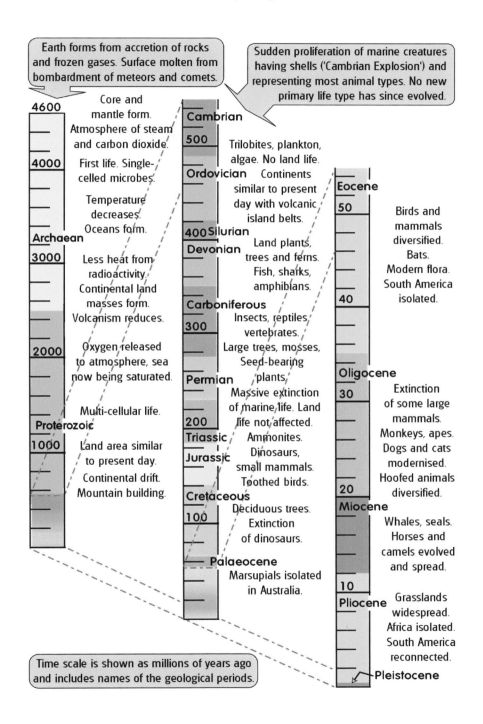

How Everything Came About

14. Earth History

15. Evolution

All life on Earth has a common origin in simple single-celled bacteria-like microorganisms that appeared about 4000 million years ago. It took a further 2000 million years or so for multi-cellular life to develop and a further 1000 million years before the first plants and animals appeared.

Since then there has been great diversification. At present, for example, there are perhaps tens of millions of animal species and a million plant species. An animal species is generally taken to be a group that will interbreed.

The species, however, fall into about 35 main groups (phyla) of animals, 12 of plants and 7 of fungi. All the main animal groups appeared between about 600 and 500 million years ago (the 'Cambrian Explosion') and, remarkably, no new major group has since evolved.

Evolution proceeds by random changes in the genes as they are passed from generation to generation. Changes that are beneficial to survival are more likely to be passed on.

16. Evolution of Humans

How Everything Came About

Million years ago

Complex organisms have gradually developed from simpler ones by genetic changes in successive generations.

PROKARYOTES (4000)
Life started with single-celled bacterial microorganisms, having a strand of DNA but no nucleus.

EUKARYOTES (1700)
The advent of the cell nucleus, with DNA within chromosomes, made multi-cellular life possible.

PLANTS (600)
Plants get energy from sunlight and store food as starch. The first plants were spore-bearing.

FUNGI

ANIMALS (600)
Animals get energy from oxidation of organic material and store food as glycogen and fat.

LAND PLANTS
First forests. (400)
Horsetails, ferns.

INVERTEBRATES

VERTEBRATES
Having backbones.

ARTHROPODS
Trilobites, arachnids, millipedes, centipedes, insects.

SPONGES (550)

BRACHIOPODS
Lamp shells.

FISHES (500)

GYMNOSPERMS
Seed-bearing (300) and able to grow in dry areas. Conifers.

ANNELIDS
Earthworms, leeches.

COELENTERATES
Corals, jelly fishes, sea anemones.

MOLLUSCS
Ammonites, oysters, clams, snails, slugs, squids, octopi.

ANGIOSPERMS
Flowering. (150)
Roses, Magnolias.

Several groups which include various less familiar sea creatures.

ECHINODERMS
Sea cucumbers, star fishes, sea urchins.

A family tree based on present knowledge, showing evolutionary paths and genetic relationships, would be enormous, yet still incomplete.

(350) **AMPHIBIANS**
Adapted to life on land, but laid eggs in water or moist places. Used skin and lungs for respiration. Frogs, toads, salamanders.

BIRDS (150)

REPTILES (350)
Eggs adapted for laying on land. Dinosaurs, alligators, turtles, lizards, snakes.

MAMMALS
Born alive. (220)

MONTREMES (70)
Hatched immature from eggs and fed on mother's milk. Platypus.

PLACENTALS (70)
Born fully developed.

(70) **MARSUPIALS**
Born immature and nourished in a pouch. Kangaroos, possums.

PRIMATES
Monkeys, (60) apes, humans.

Several groups which include elephants, horses, pigs, camels, rabbits, etc.

15. Evolution

16. Evolution of Humans

The evolution of modern humans from the rest of the animal world was a gradual process. The fossil evidence relating to the early evolution of humans is very limited, so the detailed account of events is constantly under revision.

About 10 million years ago we shared common ancestors with the apes and about 5 million years ago we were recognisable as humans.

During the next four and a half million years brain size increased, stature increased and walking became more upright.

Homo Sapiens appeared about 250,000 years ago and modern humans (Homo Sapiens Sapiens) about100,000 years ago.

Modern humans appear to have originated in southern Africa. From there they spread north and east through Europe and Asia. The Americas were colonised via a land bridge from Siberia. By 30,000 years ago modern humans were the only form of humans to be found anywhere.

How Everything Came About

Years Ago

60,000,000 — **PRIMATES**
Mammals that adapted front feet for use as hands.

There are many more species of humans that are not shown. The only surviving species of the genus Homo is Homo Sapiens..

10,000,000

6,000,000 — **APES**
Gibbons, chimpanzees, gorillas and orang-utans.

HUMANS
Upright walkers with increasingly large brains.

Several species including tree shrews, lemurs, monkeys and baboons.

4,000,000 — **AUSTRALOPITHECUS AFARENSIS**
African. Short stature (100 to 150 cm). Brain slightly larger than modern chimpanzee. Plant eaters.

PARANTHROPUS ROBUSTUS
Became extinct.

AUSTRALOPITHECUS AFRICANUS

PARANTHROPUS BOISEI
Became extinct.

3,000,000 — **HOMO HABILIS**
Taller. Hands manipulative. Shaped stone tools. Scavengers.

2,000,000 — **HOMO ERECTUS**
Taller. Walked upright efficiently. Hunted. Used fire and caves. Migrated to Asia and parts of Europe.

500,000

200,000 — **HOMO SAPIENS**
African. Used hand axes and flake tools.

NEANDERTHALS
Adapted to cold and moved to northern latitudes. Scavengers, but hunted small animals. Buried their dead and used personal decoration.

100,000 — **MODERN HUMANS**
African. Had fairly settled dwellings. Hunted large animals. Produced engravings and statues. Buried their dead. Migrated to Europe (Cro-Magnons) absorbing or supplanting Neanderthals. Migrated to other parts of the world 30,000 to 15,000 years ago. Agriculture started about 8000 years ago.

How Everything Came About

1. The First 400,000 Years

17. Elementary Particles

Energy in the form of radiation is carried by photons which have no mass. This energy can convert to mass in appropriate circumstances, and vice versa. As a consequence, there are numerous elementary particles but most are unstable, quickly changing to other particles and photons.

Ordinary matter is made up of just four particles: proton, neutron, electron and electron neutrino. The proton and the neutron are each composed of three quarks which, like the electron and electron neutrino are not divisible into smaller parts.

Particles are subject to four forces. Those with mass are subject to the gravitational force. The proton and the electron are subject to the electromagnetic force, the proton carrying a positive charge, the electron a negative charge. Like charges attract while opposite charges repel. These two forces give rise to familiar effects. The other two forces, the strong nuclear and the weak nuclear, are involved in processes within atoms.

18. Photons   21. Electrons   29. Atomic Nuclei

How Everything Came About

Quarks and leptons are the basic building blocks of matter. They have mass, spin (rotation) and a positive, negative or zero electric charge.

Light — **Medium** — **Heavy**

Quarks (charge 2/3 or -1/3, spin 1/2):
- up u^+ (light)
- charmed c^+ (medium)
- top t^+ (heavy)
- down d^- (light)
- strange s^- (medium)
- bottom b^- (heavy)

Leptons (charge -1 or 0, spin 1/2):
- electron e^{-1}
- muon μ^{-1}
- tau τ^{-1}
- electron neutrino ν_e^0
- muon neutrino ν_μ^0
- tau neutrino ν_τ^0

Antiquarks $\bar{q}$

Antiparticles are identical to normal particles except for opposite electric charge (or other charge-like feature).

Antileptons

Leptons (6, plus 6 antileptons) are stable apart from the muon and tau which decay to electrons, neutrinos and antineutrinos. Leptons and quarks are subject to the weak nuclear force. The carrier particles are the intermediate bosons.

Ordinary matter is made from these four particles.

Up and down quarks combine to form protons and neutrons, the lightest hadrons.

$u\,d\,u \rightarrow p^+$
$d\,u\,d \rightarrow n^0$

Protons and neutrons are held together by the strong nuclear force. The carrier particles are the gluons.

Protons and neutrons combine to form nuclei.

The Higgs boson is the carrier of the Higgs field, which is thought to give other particles their mass.

Electrons and nuclei combine to form atoms.

Electrons and nuclei are held together by the electromagnetic force. The carrier particle is the photon which has no mass.

Quarks can be any one of three colours, giving a total of 18, plus 18 antiquarks. They are held together by the colour force. The carrier particles of the force are gluons.

Quarks and antiquarks form hadrons. All are unstable, apart from the proton.

Combinations of three quarks form baryons, which decay to protons.

Combinations of quarks and the same number of antiquarks form mesons, which decay to electrons, photons and neutrinos.

Quarks can form other short-lived particles.

A particle and its antiparticle annihilate on meeting, releasing energy as a photon.

All particles with mass are subject to the gravitational force. The carrier particle is the graviton which has no mass and has not yet been directly observed.

17. Elementary Particles

18. Photons

Photons are familiar in constituting the radiation that is visible as light, in the range of colours from violet to red. The visible radiation is a small part of the electromagnetic spectrum which ranges from high frequency, high energy gamma and X-rays, to low frequency, low energy microwaves and radio waves.

Though not easy to appreciate, the photon behaves as both a particle and a wave motion. As a particle it has no mass but carries energy and has momentum. As a wave motion it has oscillating electric and magnetic fields.

As the carrier of the electromagnetic force, the photon is closely associated with electrically charged particles, particularly the electron. When electrons or other charged particles change their energy states, by, for example, accelerating or decelerating photons are created.

A photon of sufficient energy can transform into an electron and a positron (antielectron) which on combining revert to a photon.

The speed of a photon is, by definition, the speed of light. Because of special relativity the photon experiences no time flow. By its own clock, it arrives as soon as it has set out.

19. Light

How Everything Came About

> A photon, created when an electrically charged particle (electron, usually) accelerates or decelerates, has properties of both particle and wave. Photons are the carrier of the electromagnetic force, virtual photons passing repeatedly between electric charges.

PARTICLE
The photon has no mass, size or electric charge, but has energy and momentum.

Particle properties more evident when energy is high. ↓

WAVE
The photon has wavelength and frequency (wavelength x frequency = velocity), its frequency being a measure of its energy. The wave has an electric and a magnetic component.

Wave properties more evident when energy is low. ↓

ELECTROMAGNETIC SPECTRUM

Gamma rays	X-rays	Visible	Microwaves	Radio
0.000,000,000,1	0.000,000,1	0.000,5	10	100,000

Wavelength (mm)

Gamma rays emerge from the nucleus in nuclear reactions or in radioactivity. Their penetrating power is utilised in industrial radiography, allowing internal inspection of opaque components.

Heavy atoms bombarded with energetic electrons emit X-rays., from deceleration of the electrons, and from inner electrons raised to higher energy levels and falling back.

The penetrating power of X-rays and the wide variation in absorbing power of different materials allows the use of X-rays for industrial and medical radiography.

Outer electrons are raised to higher energy levels, by high temperatures, and then fall back, emitting visible photons.

Lasers use a cascade of electrons to obtain an intense narrow beam of photons, all in phase.

Electronic circuits force electrons to oscillate along a metal antenna and emit photons.

Electric field

Magnetic field

The process can be reversed to give oscillating electrons in circuits. This allows communications applications such as radar, radio and TV.

Microwaves heat substances throughout their bulk because their frequencies match the vibrations of atoms and molecules.

> The cosmic background radiation, which gives us a picture of the Universe 380,000 years after the Big Bang, is in the microwave band.

18. Photons

19. Light

Light is the visible part of the electromagnetic spectrum. White light is a combination of the range of visible colours, each colour corresponding to a particular frequency and wavelength.

The commonly observed and utilised effects, such as reflection, refraction, polarisation, etc., can be understood in terms of the behaviour of waves. Many equivalent effects are observed with water waves and soundwaves, for example.

At the level of individual photons, the interaction between light and the object it strikes is more complex. The photons interact with the electrons of the constituent atoms. They may raise electrons to higher energy states or eject them completely. The energised electrons will subsequently revert to more stable states, releasing photons in the process. Incident photons may recoil, losing some energy in the process, or they may destroy the electron bonding between the atoms or molecules and bring about chemical changes.

20. Lasers

How Everything Came About

> Light is a stream of photons with wavelengths in the visible range, the wavelength determining the colour that is seen. White light is a mix of all the visible colours. Photons act on the electrons of the atoms that the light strikes.

PHOTOELECTRIC EFFECT
Electron is ejected from the surface.

PHOTOCHEMISTRY
Photon breaks or makes molecular bond, as in photosynthesis.

ABSORPTION
Photon raises electron to higher energy level.

FLUORESCENCE
Immediate emission at a longer wavelength.

SCATTERING
Photon recoils giving some energy as heat to the atom.

REFLECTION
Electron falls back emitting a photon.

PHOSPHORESCENCE
Delayed emission at a longer wavelength.

TRANSMISSION

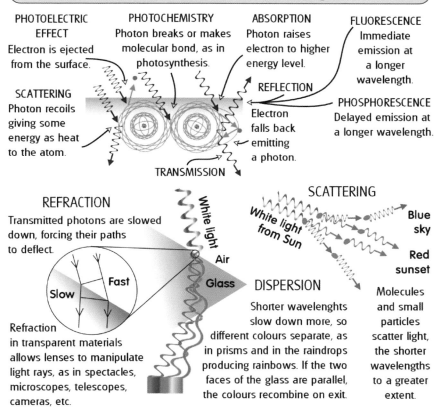

REFRACTION
Transmitted photons are slowed down, forcing their paths to deflect.

Refraction in transparent materials allows lenses to manipulate light rays, as in spectacles, microscopes, telescopes, cameras, etc.

DISPERSION
Shorter wavelenghts slow down more, so different colours separate, as in prisms and in the raindrops producing rainbows. If the two faces of the glass are parallel, the colours recombine on exit.

SCATTERING
Blue sky
Red sunset

Molecules and small particles scatter light, the shorter wavelengths to a greater extent.

POLARISED LIGHT
Light vibrates sideways in all directions. Polarising materials allow vibrations in one direction only to pass.

Crossed polarisers

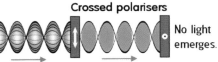

No light emerges.

Unpolarised Plane polarised

Scattered and reflected light is partly polarised, so glare is reduced by Polaroid sun glasses.

TOTAL INTERNAL REFLECTION
At grazing incidence light cannot escape from transparent materials, because the size of deflection that refraction requires is not possible. Optical fibres (glass filaments) are used in communication systems.

19. Light

20. Lasers

A laser (light amplification by stimulated emission of radiation) produces a narrow intense beam of light, but what makes it special is that all the photons are of exactly the same frequency and wavelength, and are all in phase with each other. Being in phase means that, viewed as a wave motion, the peaks and troughs coincide. Such light is said to be coherent.

Because of the precise frequency and wavelength, the laser beam does not suffer from dispersion and differential scattering but maintains its narrow width over very long distances. Being coherent the light can be used to produce interference and this has led to the development of holography. Interference occurs when coherent beams not in phase with each other are combined to give a wave pattern characterising the constituent beams.

Laser light can be produced from sources that do not have to be hot. Various crystalline, glass, liquid and gaseous materials can be used. Lasers are readily and cheaply produced and have found their way into many domestic and industrial applications.

How Everything Came About

Laser light is a narrow intense beam of photons, of the same frequency and in phase with each other. Waves are in phase when their maximum displacements coincide and their minimum displacements coincide. In-phase waves reinforce each other.

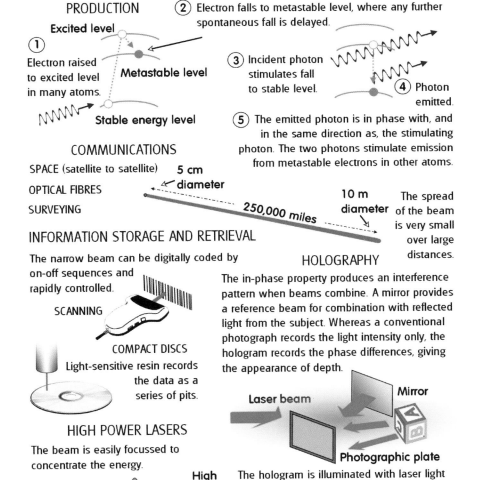

PRODUCTION

① Electron raised to excited level in many atoms.

② Electron falls to metastable level, where any further spontaneous fall is delayed.

③ Incident photon stimulates fall to stable level.

④ Photon emitted.

⑤ The emitted photon is in phase with, and in the same direction as, the stimulating photon. The two photons stimulate emission from metastable electrons in other atoms.

COMMUNICATIONS

SPACE (satellite to satellite)
OPTICAL FIBRES
SURVEYING

5 cm diameter — 250,000 miles — 10 m diameter. The spread of the beam is very small over large distances.

INFORMATION STORAGE AND RETRIEVAL

The narrow beam can be digitally coded by on-off sequences and rapidly controlled.

SCANNING

COMPACT DISCS
Light-sensitive resin records the data as a series of pits.

HIGH POWER LASERS

The beam is easily focussed to concentrate the energy.

WELDING
CUTTING
FUSION
SURGERY
WEAPONRY
MATERIALS PROCESSING

Lens — Laser beam — High Temperature spot

HOLOGRAPHY

The in-phase property produces an interference pattern when beams combine. A mirror provides a reference beam for combination with reflected light from the subject. Whereas a conventional photograph records the light intensity only, the hologram records the phase differences, giving the appearance of depth.

Laser beam — Mirror — Photographic plate

The hologram is illuminated with laser light for viewing, the interference record producing light which is an exact copy of the original reflected light from the subject. This produces a true three-dimensional effect.

Holograms for viewing in normal light consist of strips of different holograms, each from an ordinary photograh taken from a slightly different position.

17. Elementary Particles

21. Electrons

Electrons are the carriers of negative electrical charge. They are responsible for electric fields, which are regions of electrical influence, and for electric currents, which are streams of electrons moving under the influence of electric fields.

Most of the space occupied by an atom is accounted for by the cloud of electrons around the nucleus, though the nucleus represents virtually all of the mass.

Some of the electrons can be removed from the atom by friction. The separation of the negative charge of the electron and the positive charge of the nucleus gives rise to electrostatic effects. The subsequent recombination of positive and negative charges represents an electrical discharge, such as the electric shock experienced when removing garments, or the lightning seen in a thunderstorm.

An electron, like all electrically charged particles, is deflected when travelling through an electric field or a magnetic field. The electron beams that produced the pictures on early TV screens were deflected in this way to scan across and down the screen. The electron beam in an electron microscope is focussed in a similar manner.

22. Electric Currents

How Everything Came About

Electrons have mass, electric charge and spin but no size. They determine the chemical properties of atoms and their clouds around nuclei give the bulk to matter.

ELECTRIC FIELD

Around an electron (or any charged particle) is a field of influence that weakens with distance. Negative charges are repelled by the field and positive charges are attracted.

ELECTRON BEAMS

Streams of electrons are deflected by an electric field and by a magnetic field. Screen displays on oscilloscopes, early television receivers and monitors, etc., are produced this way.

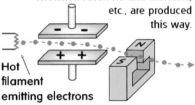

Hot filament emitting electrons

A stream of electrons (or any particles) has wave properties. Fast electrons have a shorter wavelength than light and are used in electron microscopes to give greater resolution of detail.

ELECTROSTATIC PRECIPITATOR

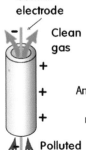

Central electrode

Clean gas

Polluted gas

Liquid droplets or solid particles are removed from gas in chemical plants, power stations, flues, air conditioning systems, hospitals, etc. An electric field in a metal tube ionises the gas molecules. The negative ions attach to the pollutants and carry them to the tube wall.

ELECTRIFICATION

Electrons can be moved by rubbing. Silk rubbed on glass picks up electrons and becomes negatively charged. The glass becomes positively charged. Metals do not exhibit static electricity as the electrons are mobile and dissipate the charge.

ELECTROSTATIC INDUCTION

Air movements transfer electrons. A negative charge on the bottom of a thunder cloud induces a positive charge on the tops of buildings, by repelling electrons. Lightning conductors attract electrons, reducing the charge above, and therefore the chance of a lightning strike. They also provide an easy path for the flow of charge in the event of a strike.

PHOTOCOPIER

A drum having a photosensitive coating is charged. An image is projected onto the drum, the light areas becoming discharged. A toner (fine black powder) is projected and adheres to the dark areas. The toner is fused to the copy by heat. Laser printers use a similar method.

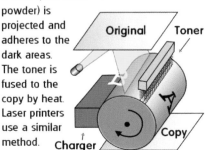

Original   Toner

Charger   Copy

45

How Everything Came About

21. Electrons

22. Electric Currents

An electric current is a moving stream of electrons. The stream moves from a region of excess electrons, negative voltage, to a region deficient in electrons, positive voltage. Between the two regions is an electric field.

Metals readily conduct electric currents because the outer electrons of the atoms are not bound to a particular atom, but can readily move from atom to atom.

Gases are normally not able to conduct electricity but, if the electric field is strong enough or the gas pressure low enough, electrons can be dragged from the atoms to produce a current.

Some liquids, electrolytes, contain molecules that split into positively and negatively charged ions when dissolved or molten. This allows conduction through the liquid via the ions, though the net effect outside the electrolyte, is a transfer of electrons.

An electric current produces a magnetic field and it usually produces heat. Certain materials exhibit superconductivity, at very low temperatures. Currents once established in these materials flow indefinitely, without loss of energy and therefore without production of heat.

23. Magnetism

How Everything Came About

An electric current is a flow of electrons under the influence of an electric field, the field arising between regions of relative electron deficiency (positive) and surplus (negative).

METALS

Metals conduct electricity readily because the outer electrons are not tied to individual atoms but can move through the crystal lattice.

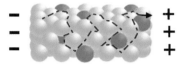

Though the electrical impulse moves at the speed of light, the electrons progress slowly. They are deflected by impurity atoms, lattice defects and thermal vibrations of the atoms. Their energy appears as electrical resistance and, subsequently, heat. The best conductors are pure metals or alloys at low temperature.

HEATING EFFECT

Resistance results in heat, which in turn causes a resistance increase. The heating is used in light bulbs, ovens, furnaces, etc.

MAGNETIC EFFECT

When an electric current flows, a magnetic field is produced, its lines of force being at right angles to the current.

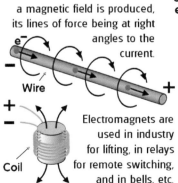

Electromagnets are used in industry for lifting, in relays for remote switching, and in bells, etc.

GASES

Gases conduct electricity if the field is strong enough or the pressure low enough. Atoms become positively charged ions as electrons are removed by the field.

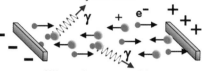

When electrons and ions recombine, photons are emitted, as in sodium vapour, mercury vapour, arc, neon and fluorescent lamps.

LIQUIDS

Electrolytes (molten salts and most solutions of inorganic compounds) pass current by ion movement, and decompose in the process (electrolysis).

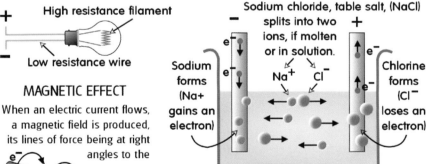

Sodium chloride, table salt, (NaCl) splits into two ions, if molten or in solution. Sodium forms (Na+ gains an electron). Chlorine forms (Cl⁻ loses an electron).

Electrolysis is used for production of chemicals, and extraction, electroplating and polishing of metals. Some corrosion processes involve electrolysis.

SUPERCONDUCTIVITY

At very low temperatures, in many materials, the electron motion becomes ordered and there is no resistance. The effect is used for magnets, generators, motors, cables and levitated trains.

22. Electric Currents

23. Magnetism

An electron has spin and the resulting motion of its electric charge produces a magnetic field.

Depending on the number and arrangement of the electrons in the atom, a material can exhibit different degrees of magnetic properties. The strongly magnetic materials, iron, cobalt and nickel, have atoms with a strong net magnetic field. This leads to alignment of the atoms in the material when an external magnetic field is present, that is, magnetisation.

A magnetic field is a region of influence between a magnetic north pole and a magnetic south pole. Neither pole can exist separately. This contrasts with an electric field which surrounds an electric charge. The charge can be positive or negative and each type of charge can have separate existence.

The nuclei of some atoms have magnetic fields when the proton and neutron spins do not cancel out. This does not give rise to any commonly encountered effects, but it has led to the development of nuclear magnetic resonance (NMR), which is a method of imaging the internal structure of materials. One application is magnetic resonance imaging (MRI) which is used routinely for imaging the human body for medical purposes without any danger of tissue damage.

24. Electric Power

How Everything Came About

FERROMAGNETISM

Magnetic field
Electron spin
Atomic fields line up

An electron has spin and therefore a magnetic field. In atoms, the electrons in full electron shells form pairs with opposite spin, but unfilled shells tend to have electrons with parallel spin and hence a net magnetic field. Iron has five parallel spin electrons.

The atoms line up in regions (domains) to balance their fields. Application of an external field moves domain walls and, if strong enough, rotates the fields within the domains. Iron, nickel, cobalt and some alloys are ferromagnetic. The effect disappears at the Curie temperature, thermal agitation then destroying the orientation.

Domain
No external field
Weak external field
Strong external field

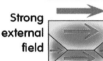

PARAMAGNETISM

All atoms and molecules with an odd number of electrons, metals, some atoms having unfilled inner electron shells, and some other compounds including molecular oxygen are paramagnetic. The atom or molecule has a net magnetic field. The material can be magnetised slightly.

A moving electric charge produces a magnetic field.

MAGNETS

Every magnet has two poles, north and south, between which is a magnetic field. An isolated pole cannot exist. (Electric fields differ: single charges, positive or negative, exist and have radiating fields.) Permanent magnets ('hard') are of alloys that retain magnetisation well. Crystal imperfections and boundaries hamper movement of the domain walls. Electromagnets, of iron ('soft'), lose magnetisation easily.

DIAMAGNETISM

Nearly all molecules with an even number of electrons, and most non-metallic solids, are diamagnetic. The internal magnetic fields cancel out. The material is repelled by a magnet but the effect is very small.

MAGNETOSTRICTION

Ferromagnetic materials generally lengthen slightly when magnetised. The effect is utilised in strain transducers.

NUCLEAR MAGNETIC RESONANCE

Depending on the relative numbers of protons and neutrons, some nuclei have a magnetic field. If an external magnetic field is applied, the nuclei can absorb and then emit radio frequency radiation. Pulses of radiowaves are used to excite the nuclei and the resulting emission is related to the strength of the applied magnetic field and the identity of the nuclei. By using gradients in the magnetic field, particular nuclei can be located and mapped. The technique is used in studies of materials. Magnetic resonance imaging (MRI) is an application used routinely for medically imaging the human body. It is superior to X-ray tomography and it does not damage tissue as X-rays do..

23. Magnetism

24. Electric Power

The generator is the major source of electrical power. It is based on the principle that, just as an electric current always produces a magnetic field, so a conducting material moving through a magnetic field will have a current induced in it. The reverse effect, an induced motion of a current-carrying conductor placed in a magnetic field, is the basis of electric motors. An electric motor converts the electrical energy into mechanical energy.

Alternating current in which the electron flow repeatedly reverses direction is more easily generated and more easily employed in motors. Moreover its voltage (the electrical pressure arising from the field between positive and negative terminals) can be easily increased or decreased by transformers. The distribution of power from generating stations is in the form of alternating current.

Electric cells and batteries produce direct current chemically by having electrodes of appropriate materials immersed in electrolytes.

In fuel cells the electrodes act as catalysts and are not altered by the reactions. Usually hydrogen is combined with oxygen, water being the waste product.

In solar cells electrons are liberated by direct impingement of photons

25. Electronic Systems

CELLS AND BATTERIES

Electrode materials and electrolytes are chosen so that in a chemical process electrons are drawn from one electrode and fed to the other. In rechargeable batteries the process can be reversed by electrolysis.

In a car battery, sulphuric acid (H_2SO_4) is consumed, producing lead sulphate and water.

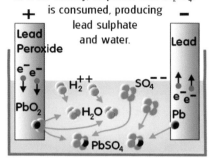

FUEL CELL

Electrodes that are catalysts are not consumed, so continuous production of electricity is possible. Usually hydrogen and oxygen are combined, giving water. Porous electrodes provide a large surface area of contact with the gas.

GENERATOR

If a conductor cuts through a magnetic field, electrons flow at right angles to the field and to the direction of motion. Most electrical power is generated by rotating a coil in a magnetic field. Motors use the reverse effect.

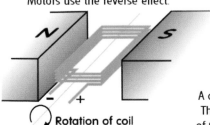

Rotation of coil

SOLAR CELLS

Some semiconductors produce electron flow when photons impinge (photovoltaic effect). This is the basis of the solar cell as used in watches, calculators, etc. and solar panels for space vehicles and heating of buildings.

PIEZOELECTRICITY

Many non-conducting crystals, when distorted, develop charged regions as constituent particles are moved from their equilibrium positions. Sound waves distort the crystals, as in microphones and vibration sensors. Loudspeakers, headphones, sonar, asdic and ultrasonic cleaning use the reverse effect.

THERMOELECTRICITY

If two different conducting materials are joined in a loop, with one junction hot and the other cold, a current flows, because the heat flow is partly movement of electrons. The reverse effect is used for heaters and coolers.

AC/DC

Because the two sides of a generator coil pass through the field in alternate directions, the current repeatedly changes direction. This gives alternating current (AC). Batteries, etc,. give direct current (DC).

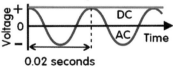

0.02 seconds

With AC, the voltage (the electrical pressure driving the current) can easily be increased or decreased by a transformer.

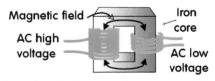

A coil carrying AC gives a varying magnetic field. This produces AC in a second coil. The number of turns in each coil controls the voltage change.

24. Electric Power

25. Electronic Systems

A transducer is a device for accepting a signal in one form of energy and converting it to an alternative form. A microphone, for example, converts a sound signal to an electrical signal. A loudspeaker does the reverse, converting an electrical signal to a sound signal.

Because electrical signals can be readily and rapidly processed and transmitted, most transducers are used to convert to and from electrical signals.

This has given rise to a vast number of electronic systems used in domestic and industrial applications. Most domestic applications are in the fields of control of machines (washing machines, cars, etc.), entertainment (CD and DVD players, etc.), and communication (radio, telephones, mobile phones, etc.).

The description 'electronic' as opposed to 'electrical' implies that the electricity employed is carrying signals rather than transmitting power.

26. Electronic Components

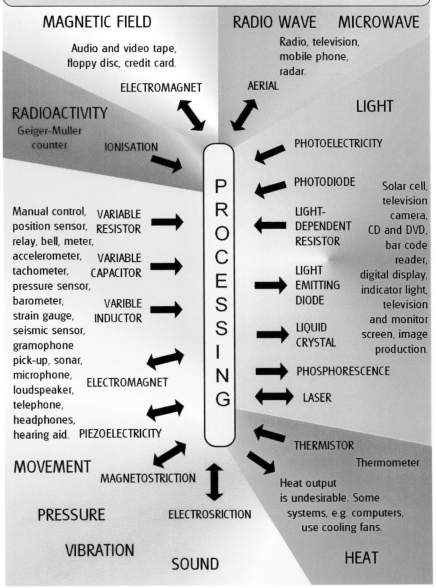

25. Electronic Systems

26. Electronic Components

Electronic systems are complex in design but based on relatively few kinds of components. Each component regulates in some way the flow of current.

The resistor reduces current flow whether it be direct or alternating. The capacitor prevents the flow of direct current but conducts alternating current, more readily the higher the frequency. The inductor allows the flow of direct current, and conducts alternating current, more readily the lower the frequency.

Amplification of signals is a key process in electronic circuits and this is achieved by the transistor which can also act as a switch.

Prior to the invention of the transistor, high voltage was necessary to power electronic circuits and this severely restricted the development of portable equipment The transistor was a major breakthrough, but the integrated circuit (chip), which can currently incorporate billions of transistors, had an even greater impact.

Logic gates are vital for the processing of numerical data. Assemblies of transistors are arranged so that the output is determined by the combination of inputs. The gates act as switches: the inputs and outputs are either on or off and the size of the signal is not relevant.

27. Electronic Processing

How Everything Came About

PRINTED CIRCUIT BOARD (PCB)

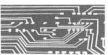

Carries current and voltage between components, in place of wiring. Consists of an insulating board with preformed solder tracks on which the circuit components are mounted.

RESISTOR

Limits current flow and reduces voltage, AC or DC. Consists of a thin film of carbon or metal, or a length of fine wire.

CAPACITOR

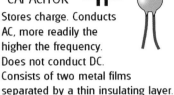

Stores charge. Conducts AC, more readily the higher the frequency. Does not conduct DC. Consists of two metal films separated by a thin insulating layer.

INDUCTOR

Conducts AC, more readily the lower the frequency. Conducts DC. Consists of a coil of wire, with or without a central rod of magnetic material.

Traditional electronic components vary in size and shape. Symbols used in circuit diagrams are shown with typical examples.

SEMICONDUCTOR DEVICES

JUNCTION DIODE

Conducts current in one direction only.

JUNCTION TRANSISTOR

Amplifies. Acts as a switch.

INTEGRATED CIRCUIT (IC, CHIP)

The traditional PCB with its circuit components has been largely replaced by integrated circuits, particularly where small size or much processing are required. Modern integrated circuits can comprise billions of transistors and associated components.

OPERATIONAL AMPLIFIER

Integrated circuit used in analogue processing, giving a high level of amplification and mixing or adding of inputs.

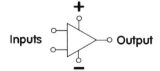

Inputs → Output

LOGIC GATE
Integrated circuit used in digital processing, giving high (1) or low (0) output depending on input.

INPUT	A	0	0	1	1
	B	0	1	0	1
OUTPUT	OR	0	1	1	1
	NOR	1	0	0	0
	AND	0	0	0	1
	NAND	1	1	1	0
	NOT		1		0

26. Electronic Components

27. Electronic Processing

Two different kinds of signal, analogue and digital, are processed electronically.

An analogue signal is one that varies continuously with time, such as a sound or a changing temperature. Analogue electrical signals can be modified and amplified to drive output transducers, as in radios, alarms, etc.

A problem with analogue signals is that they can easily become distorted in transmission and pick up extraneous background effects. This has led to greater use of digital signals.

A digital signal is a sequence of states, each one being on or off. The actual size of the signal during the on periods is not relevant. Analogue signals are converted to digital signals by a very rapid sampling of the signal and representing each value by is value expressed in binary notation. Binary notation of numbers requires only two digits, 0 and 1, and each number is thus equivalent to a digital sequence.

Optical processing using the photon rather than the electron as the carrier, offers advantages. The photon has no mass and generates very little heat in transit.

28. Semiconductors

How Everything Came About

Electrons move slowly through conductors, but the electrical impulse travels at the speed of light and requires only a low power level to drive it. This has made electric currents suitable for processing information. (The photon is even better suited, hence the progress in fibre optics and optical processing.)

ANALOGUE INFORMATION
Analogue information varies continuously with time.

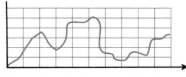

Processing involves amplification, enhancing the useful data and removing spurious signals (noise) or unwanted data. Oscillator circuits produce signals of particlar frequencies and accurate timing pulses.

RADIO WAVES
Radiowaves provide carrier waves for the information. The carrier wave allows the receiver to 'tune in' to the required signal and is then removed.

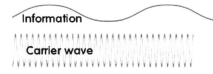

In amplitude modulation (AM) the sizes of the two waves are combined. In frequency modulation (FM) the two frequencies are combined.

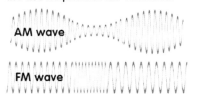

DIGITAL INFORMATION
Digital information consists of a sequence of on or off pulses and is not affected by signal distortion.

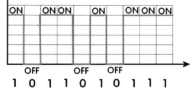

1 0 1 1 0 1 0 1 1 1

The pulses are interpreted as binary numbers and converted to decimal values as required.

1011010111 (Binary) = 727 (Decimal)

Switching circuits mimic binary numbers by giving outputs high (on) or low (off). The circuits are combined in different ways in logic gates. These perform arithmetic operations in, e.g. calculators and computers.

BINARY NUMBERS
0 = 0
1 = 1
10 = 2
11 = 3
100 = 4
101 = 5
110 = 6
111 = 7
1000 = 8
etc.

ANALOGUE TO DIGITAL CONVERSION
Analogue signals are sampled very rapidly to give a series of values. Each value is converted to a binary number.

MULTIPLEXING
Several signals can be transmitted at the same time along one channel. Analogue signals each have a different carrier frequency. Digital signals share on a time basis.

27. Electronic Processing

28. Semiconductors

Apart from metals and insulators, there is a third category of materials, the less familiar semiconductors. These can conduct electricity, but weakly compared with metals. Their conduction arises from a limited number of electrons that break away from their atoms, because of thermal agitation.

By adding certain impurities in very small quantities the conduction can be increased. The choice of impurities is based on the electron dispositions. In the n-type semiconductor, electrons carry the current. In the p-type, there is a deficit of electrons which therefore move from vacancy to vacancy giving the effect of conduction by positively charged holes.

The importance of semiconductors lies in the invention of the transistor. By using n- and p-types an electronic valve was produced which allowed a current flow to be controlled by a small applied voltage. Thus switching and amplification could be achieved.

Other specialised devices are based on semiconductors and have the advantage of low power consumption.

INTRINSIC SEMICONDUCTORS

Some covalently bonded elements and compounds conduct electricity, very weakly compared with metals. Thermal vibrations cause some electrons to break away and carry a small current. Thus, conduction increases as temperature rises, unlke metals.

> The importance of semiconductors lies in the electronic devices that have been developed from them, notably the transistor.

EXTRINSIC SEMICONDUCTORS

By adding small amounts of certain impurities conduction is increased.

n-TYPE (DONOR IMPURITIES)

If the impurity atom has an electron more in its outer incomplete shell, the electron is not required for bonding and is available for conduction. Silicon doped with arsenic is an example.

p-TYPE (ACCEPTOR IMPURITIES)

If the impurity atom has an electron fewer, a vacancy or hole results. Conduction increases by hole movement, i.e. an electron jumps into the hole creating a further hole, and so on. Silicon doped with gallium is an example.

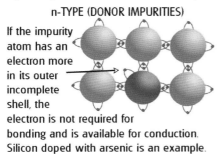
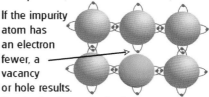

THERMISTOR

Temperature is measured by the change in resistance.

LIGHT-DEPENDENT RESISTOR (LDR)

Photons impinging on some semiconductors break bonds and release electrons for conduction.

p-n JUNCTION DIODE

A single crystal is produced, n-type at one end, p-type at the other. Conduction occurs in one direction only. With forward bias, electrons and holes combine, so current flows. With reverse bias, electrons and holes separate and cannot combine.

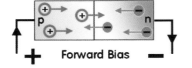
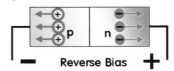

n-p-n JUNCTION TRANSISTOR

A voltage on the thin base layer starts a flow of current, but most of the moving electrons are drawn to the positively charged collector. Thus a switching effect and a large amplification of current results.

LIGHT-EMITTING DIODE (LED)

If the n-p junction is near the surface, photons are emitted when electrons and holes combine.

PHOTODIODE

With reverse bias, impinging photons can break bonds and release electrons and holes for conduction.

17. Elementary Particles

29. Atomic Nuclei

The periodic table is a list of all possible elements that exist, an element being a substance consisting of atoms all having the same number of protons in their nuclei. The elements are arranged in order of their proton number. Thus, hydrogen has one proton, helium has two, lithium has three, and so on, up to uranium which has 92. Elements beyond uranium have been produced but are unstable and decay rapidly.

The elements are distinguished as different substances because the number of protons in the nucleus determines the number of electrons that the atom will have which in turn determines the chemical properties of the substance.

The atomic nucleus contains neutrons in addition to protons and the number of neutrons is roughly the same as the proton number. Different isotopes of the same element have a slightly different number of neutrons. Chemically, isotopes of the same element cannot be distinguished, though the mass is of course, slightly different. The mass of the nucleus is denoted by the mass number, which is the proton number plus the neutron number.

30. Radioactivity   32. Atoms

How Everything Came About

Hydrogen has the simplest nucleus, consisting of one proton. Additional protons and neutrons build up the periodic table. Adding a proton makes a new element, and adding a neutron makes a different isotope of the same element.

The proton or neutron diameter is about 0.000,000,000,002 mm.

The density of the nucleus is very great: a cubic millimetre weighs over 200,000,000 kilograms.

The size and weight of the nucleus increases in proportion to the number of protons and neutrons.

The chart shows the composition of the stable nuclei and some naturally occuring unstable nuclei.

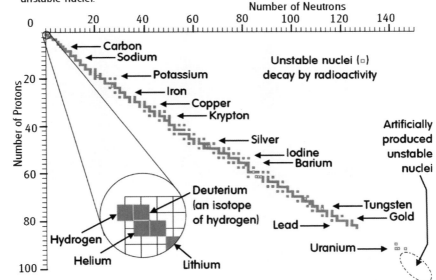

Why are there gaps in the sequence as protons and neutrons are added?	Why are there roughly equal numbers of protons and neutrons in the lighter nuclei?	Why is there an excess of neutrons in the heavier nuclei?
Protons and neutrons occupy sequentially filled energy levels. Two protons, or neutrons, with opposite spin, can occupy each proton, or neutron, energy level. Hence stability favours even numbers of protons and even numbers of neutrons.	A neutron can decay to a proton, to occupy a lower energy level, so the proton and neutron levels tend to fill equally. Hence stability favours equal numbers of protons and neutrons.	As proton numbers increase, electric repulsion becomes important in comparison with the short-range strong nuclear attractive force. Neutrons have no electric charge. Hence stability begins to favour fewer protons.

29. Atomic Nuclei

30. Radioactivity

Some atomic nuclei are unstable and spontaneously emit a particle. A helium nucleus, i.e. two protons and two neutrons, may be emitted (alpha decay). A neutron may change to a proton, electron and antineutrino, emitting the latter two particles (beta decay). A high energy photon may be emitted (gamma decay). In alpha and beta decay, because the number of protons is changed, the element changes from one substance to a different one.

Radioactive dating makes use of decay rates that lie within the time periods of interest. In carbon dating the age of an object is indicated by the proportion of carbon 14 present. Carbon 14 is an isotope of carbon having 6 protons and 8 neutrons, which changes by beta decay to nitrogen 14, which has 7 protons and 7 neutrons. The proportion of the radioactive isotope present when carbon was incorporated in the item is known from the proportion maintained in the atmosphere.

Radiation can be hazardous in damaging living tissue, but is hence useful for sterilisation.

The ability to trace radioactive isotopes by detecting the emitted particles allows numerous industrial and medical applications. Positron emission tomography (PET) is an important method of scanning in medical diagnosis.

31. Nuclear Energy

How Everything Came About

Some naturally occuring nuclei and all artificially produced ones are unstable. They spontaneously emit one or more particles. The changed nucleus may be stable or not.

ALPHA (α) DECAY

 He

A helium nucleus is emitted, carrying away 2 protons. The element moves two places down the periodic table.

BETA (β) DECAY

→ • e^-
→ • $\bar{\nu}$

A neutron decays to a proton, creating an electron and an antineutrino. The element moves one place up the periodic table.

GAMMA (γ) DECAY

 γ

A photon is emitted when a nucleus falls to a lower energy level. The element remains the same.

PENETRATION

A few cm of air — **Thin sheet metal** — **Several cm of lead**

Medical and industrial radiography applications, and sheet thickness gauges, utilise the ability of radiation to penetrate opaque materials.

HALF-LIFE

The decay of each nucleus is random and not influenced by any process. The half-life is the time after which half of the nuclei in any quantity will have decayed.

CHEMICAL EFFECTS

Radiation damages living cells but is used for medical therapy and for sterilization. Gamma rays present serious hazards though alpha emitters are hazardous only if ingested. Radiation is used to modify plastics.

TRACING

Radioactive isotopes have identical chemical properties to other isotopes of the same element. Thus, movement of elements through chemical, physical and biological steps can be traced, in medical diagnosis and industrial processes

POSITRON EMISSION

Some artificially produced radioactive nuclei decay in other ways. Of importance is positron emission, used in a type of medical tomography scan (PET).
A proton converts to a neutron, emitting a positron and a neutrino.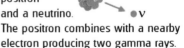
The positron combines with a nearby electron producing two gamma rays.

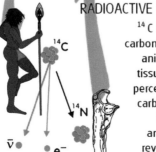

RADIOACTIVE DATING

^{14}C (an isotope of carbon) is taken into animal and plant tissue at a known percentage of total carbon. The lower level of ^{14}C in ancient remains reveals their age.

Heat from radioactivity in the Earth is responsible for plate movement, mountain building, earthquakes, etc.

30. Radioactivity

31. Nuclear Energy

Iron which has 26 protons is the most stable element. Elements with fewer protons will release energy, if combined to make heavier elements. Elements with more protons than iron will release energy, if split to make lighter elements. These two processes, fusion and fission, are the means of obtaining nuclear energy.

The energy released comes from the loss of a small proportion of the mass of the nuclei involved.

Many nuclei must take part in the process to give a useful amount of released energy. For fission, a chain reaction is employed. The uranium nucleus splits when it accepts a neutron. It then releases further neutrons which cause neighbouring nuclei to split and so on. In an atom bomb the process is uncontrolled. In a fission reactor, control rods absorb neutrons to regulate the process.

The fusion process has been used in an uncontrolled way in the hydrogen bomb. A controlled continuous fusion process has still not been achieved for an appreciable length of time. The very high temperature of the plasma of nuclei creates confinement and other difficulties.

The energy radiated by stars is produced by fusion.

How Everything Came About

FUSION   PERIODIC TABLE OF THE ELEMENTS   FISSION

1	2	3
H	He	Li

25	26	27
Mn	Fe	Co

91	92
Pa	U

If nuclei of light elements combine to make heavier ones, the total mass decreases.

Iron is the most stable element.

If nuclei of heavy elements fragment to make lighter ones, the total mass decreases.

The lost mass appears as energy in the form of photons according to $E = mc^2$, where c is the speed of light in vacuum. One gram of matter yields 25 million kilowatt hours.

FUSION REACTION

Because nuclei are positively charged, high temperatures, (100,000,000 °C), are needed to bring them close enough to combine. Two ^{2}H (deuterium, an isotope of hydrogen) nuclei combine to give ^{3}He (an isotope of helium) and a neutron.

HYDROGEN BOMB

Deuterium gas is raised to 350,000,000 °C by a fission bomb to produce the fusion reaction.

FISSION REACTOR

Over 50 years of effort has not yet made fission commercially viable. Confinement of the hot plasma is difficult. Magnetic fields, which deflect the moving charged nuclei, keep them from the wall of the toroidal container.

Magnetic fields

FISSION REACTION

When ^{235}U, an isotope of uranium, captures a neutron, it splits and emits further neutrons.

ATOMIC BOMB

The ^{235}U is initially shaped to allow leakage of neutrons. A conventional explosion shapes the ^{235}U to reduce the surface area and the chain reaction proceeds.

The amount of ^{235}U has to be greater than a critical mass or neutron leakage would still be too great.

NUCLEAR REACTOR

The number of neutrons producing further fission has to be carefully limited. Control rods that absorb neutrons are raised and lowered amid the fuel rods.

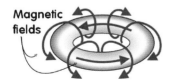

FUSION IN STARS

Hydrogen is converted to helium by fusion. Helium nuclei combine, in threes, to form carbon, and subsequent additions of helium account for alternate heavier members of the periodic table. In-between nuclei are formed by radioactive decay. The process cannot progress beyond iron: heavier nuclei are produced in supernova explosions.

29. Atomic Nuclei

32. Atoms

An atom nucleus has a number of protons and a similar number of neutrons. Each proton has a positive charge which can attract and hold in position an electron. Each nucleus therefore has around it a number of electrons equal to the number of protons. This makes up an atom of the appropriate element.

The electrons are visualised as orbiting the nucleus but a more realistic picture is that the electrons exist as a cloud, which expresses the probability of the electron being in a particular place at a particular time.

As more electrons are added for the heavier nuclei, they are located further from the nucleus and therefore have different energy levels. Only two electrons, one with spin up and one with spin down (clockwise and anticlockwise), can occupy the same energy level. Groups of energy levels form shells and subshells which represent particularly stable arrangements.

The chemical properties of the different elements depend on the configurations of the outer electrons.

The size of an atom is determined by the electron cloud and is much bigger than the nucleus. Thus an atom is mostly empty space.

33. Molecules

How Everything Came About

Atoms consist of nuclei surrounded by electrons. The electrons, negatively charged, are held by the attraction of the positively charged protons in the nucleus. The atom has an equal number of electrons and protons.

The atom is about 10,000 times as big as the nucleus and cannot be shown to scale.

The electrons can be visualised as orbiting the nucleus. In reality, there is a 'cloud' within which each electron has a probability of being at a specified location.

Because the energy and spin of the electrons are quantised, there are discrete energy levels that the electrons occupy. Only two electrons, with opposite spin, can occupy each energy level.

IONS
An ion is an atom that has one or more electrons extra, or is one or more electrons short. It thus has a negative or positive charge. Electrons are often included in the term 'negative ions'.

The energy levels are grouped in shells and subshells, lying at different distances from the nucleus and representing particularly stable arrangements.

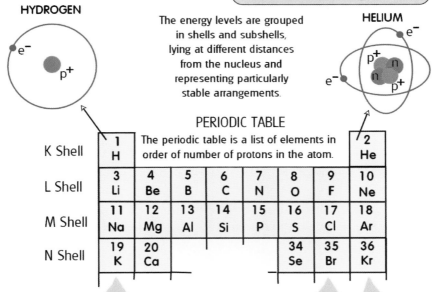

PERIODIC TABLE

The periodic table is a list of elements in order of number of protons in the atom.

K Shell	1 H							2 He
L Shell	3 Li	4 Be	5 B	6 C	7 N	8 O	9 F	10 Ne
M Shell	11 Na	12 Mg	13 Al	14 Si	15 P	16 S	17 Cl	18 Ar
N Shell	19 K	20 Ca				34 Se	35 Br	36 Kr

Atoms with complete shells plus one extra electron

Atoms that are one electron short of a complete shell

Atoms with complete shells form elements that are particularly stable and unreactive.

Elements with the same number of electrons additional to complete shells have similar chemical properties, as do those with the same number of electrons short of complete shells.

32. Atoms

33. Molecules

Atoms bond together to make molecules. Atoms that bond to their own kind can account for only 92 substances: most of the substances that exist are made up of more than one element. Such substances are compounds. Some form crystalline structures in which the molecule repeats indefinitely in a geometrical lattice. Some form long chains of repeating molecules.

Molecules that are based on carbon are called organic because biological material is composed largely of such molecules. The number of possible organic molecules is enormous because carbon has the ability to bond with itself indefinitely, creating very large molecules of great variety.

The bonding between the atoms in molecules is due to the interaction between the outer electrons of neighbouring atoms.

A weaker form of bonding holds molecules together. Molecules have positive and negative regions which attract, respectively, the negative and positive regions of neighbouring molecules.

A proper description of the bonding is provided by quantum theory but the calculations are difficult for all but the simplest structures.

34. Sound   35. Heat and Temperature   36. Liquids
44. Chemical Processes

How Everything Came About

Atoms bond together by the interaction of the electrons that are surplus to complete shells. Bonded atoms form molecules but some structures continue indefinitely and individual molecules cannot be distinguished.

PRIMARY BONDS

There are three types of primary bond but bonds can have mixed features.

COVALENT BOND
Electrons are shared between two or more atoms to complete their shells.

IONIC BOND
One or more electrons are transferred to give both atoms complete shells.

A positive and a negative ion are formed and attract each other.

METALLIC BOND
Loosely bonded electrons move easily through a lattice of atoms, binding the atoms together.

 Cu

It is not possible to distinguish single molecules. The atoms form simple closely packed crystal lattices.

Because the incomplete shells are not symmetrical, the bonds are directional.

H_2O Water molecule

Some molecules form crystal lattices, often complex because of the directional bonding. Some molecules are chain-like and can be very large.

 NaCl

Sodium chloride (table salt) molecule

The attraction is equal in all directions, so a simple lattice is formed, the symmetry depending on the sizes of the atoms.

NaCl crystal

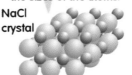

Copper crystal

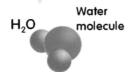

Molecules based on carbon are classed as organic because many of them constitute biological material.

Compounds are substances consisting of molecules having atoms of more than one element.

SECONDARY BONDS

Molecules are held together in substances by secondary bonds. The bonds are between positve and negative regions of neighbouring molecules.

DIPOLE BOND
H_2O

VAN DER WAALS BOND

Polar molecules are those with permanent positive and negative regions. This allows each molecule to bond to its neighbours.

Electron motion causes fluctuating positive and negative regions in all molecules.

33. Molecules

34. Sound

Sound is a wave motion that can travel through solids, liquids and gases. Its origin is a vibration of an assembly of molecules which sets adjacent molecules vibrating which, in turn, set others vibrating. This chain of push-pull motion constitutes a wave with frequency and wavelength. Generally, a sound will be made up of waves of various frequencies.

Only a limited range of frequencies is audible to the human ear. Musical notes are audible sounds of particular frequency produced by the vocal chords or musical instruments. The same note played on different instruments sounds different because of overtones. These are extraneous vibrations with frequencies numerically related to the fundamental frequency.

The penetrating ability of sound waves is utilised in applications of ultrasonics, such as sonar and medical diagnosis and treatment.

The Doppler Effect is a commonly observed phenomenon when a siren or whistle on a moving vehicle is heard to change frequency as the vehicle passes.

Sonic booms occur because the sound waves cannot travel fast enough to move ahead of the supersonic aircraft.

How Everything Came About

Sound is a wave motion that propagates through materials by a to-and-fro vibration of molecules. Collisions between molecules transmit the motion. Sound travels faster in liquids and solids than in gases as the molecules are not so far apart. It cannot travel through empty space.

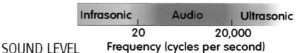

SOUND LEVEL

Frequency (cycles per second)

Barely audible sound is 0 decibels (dB), conversation is about 60 dB and pain is felt at about 120 dB. An increase of 10 dB represents a ten-fold increase in intensity (energy).

MUSIC

Musical notes are sounds of particular frequency (speed divided by wavelength). Scales are sequences of notes related by fixed frequency ratios. An octave represents a doubling of frequency, for example middle C is 264 cycles per second, top C is 528.

PRODUCTION OF NOTES

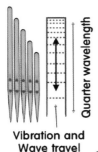

Vibration and Wave travel

Musical instruments set up standing waves by reflecting a wave back and forth along the same path. The path may be a column of air (wind), a tensioned wire (string) or various membranes or metal shapes (percussion). The path length determines the wavelength of the note and hence the frequency (pitch).

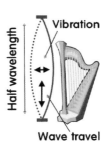

Wave travel

OVERTONES

Vibrations of higher frequencies accompany the fundamental frequency, because the path length can fit a range of shorter wavelengths. Different instruments playing the same note sound different, because of the presence of overtones at different sound levels. Fundamentals and overtones constitute harmonics.

DOPPLER EFFECT

When a sound source is approaching, the frequency heard is higher than when it is receding, because the crests of the waves are emitted closer together as the source advances.

ULTRASONICS

High frequency inaudible waves are used in sonar for detection and ranging underwater. Bats use a similar system for navigation. Ultrasonics are also used for medical diagnosis and treatment, and detecting flaws in manufactured goods.

SHOCK WAVES

Sound wave crests cannot move away from an aircraft travelling at supersonic speeds. They pile up as a large crest, forming a cone which extends backwards and can cause damage as it traces a path across the ground.

33. Molecules

35. Heat and Temperature

Atoms and molecules are constantly in motion. In solids the molecules vibrate. In liquids and gases the molecules can also move around.

Heat is this energy of motion in transit from place to place. Temperature is a measure of the level of heat, meaning that if heat is transferred from one object to another, we deem the recipient to be at a lower temperature than the donor. There is no heat transfer between bodies at the same temperature.

Heat transfer can occur in three ways, conduction, convection and radiation.

Conduction occurs between bodies in contact. Heat is transferred by the collisions between mobile or vibrating molecules.

Convection occurs in gases and liquids. Warmer zones, having greater molecular motion, expand and rise, while cooler zones descend.

Everything emits (and receives) heat by electromagnetic radiation, mainly in the infrared frequencies. The amount emitted increases with the temperature of the emitting surface.

As temperature is increased the bonds between molecules are broken. Solids melt to liquids which in turn boil. At very high temperatures, electrons are stripped from the atomic nuclei to form plasmas.

How Everything Came About

Atoms and molecules are in motion, the energy appearing as heat. Temperature is a measure of the level of the energy. As the energy increases, molecular bonds are broken and solids melt, vaporise and (at very high temperatures) become plasmas.

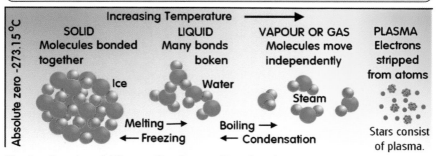

Absolute zero -273.15 °C

Increasing Temperature →

SOLID — Molecules bonded together (Ice)

LIQUID — Many bonds boken (Water)

Melting → ← Freezing

VAPOUR OR GAS — Molecules move independently (Steam)

Boiling → ← Condensation

PLASMA — Electrons stripped from atoms. Stars consist of plasma.

The bonding gives rigidity. Crystalline materials have precise melting points. Amorphous materials soften progressively.

CONDUCTION
Heat flow transfers energy from hot bodies to cold, by molecular collisions in liquids and gases, by vibrations in non-conductors of electricity, and by vibrations and electron flow in conducting solids.

REFRIGERATION
Heat can be moved from cooler to hotter bodies, but only by expending additional energy. A liquid with low boiling point is pumped through a restriction. The drop in pressure causes the liquid to vaporise and absorb heat.

The degree of bonding gives resistance to flow (viscosity) and droplet and bubble formation (surface tension).

LATENT HEAT
(of Fusion and Vaporisation)
On melting or boiling, extra heat is needed to break the molecular bonds. No temperature rise occurs.

EXPANSION
Heat causes expansion because of the increased motion, but some substances, e.g. water, expand on freezing. Thermostats use the expansion to control appliances. Gaps are provided in engineering structures to allow for expansion.

CONVECTION
Expansion decreases density, so hotter regions of liquids and gases rise while cooler ones descend.

Gas pressure is caused by molecular collisions. Changes in pressure affect boiling and melting temperatures. Vapours are gases that can be compressed back to liquids without first reducing the temperature.

Absolute zero, when atoms and molecules have the lowest possible energy, is the lowest temperature that could ever be reached.

RADIATION
Heat raises outer electrons to higher energy levels. They fall back emitting photons, mainly of infrared wavelenghts. Hot bodies radiate more than they receive and thereby cool.

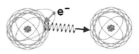

33. Molecules

36. Liquids

In liquids, the vibrational energy of the molecules is high enough to break bonds and allow molecules to move around. The bonding is still sufficient to hold the substance together in a fixed volume though some molecules that gain, by chance, sufficient energy do escape. This evaporation causes cooling because the escaping molecules carry higher than average energy.

The degree of the bonding in liquids is exhibited in surface tension, wetting or non-wetting of surfaces and in viscosity.

The limited nature of the bonding is exhibited in the way a liquid fills a volume of any shape, and the way pressure is transmitted uniformly through the liquid in all directions.

Most liquids encountered at normal temperatures will mix with water or with oil. Emulsifying agents allow the two sorts to be dispersed in each other.

37. Liquid Crystals

How Everything Came About

Liquid structures are intermediate between those of solids and gases. The molecules have sufficient thermal energy to move with respect to each other but the bonding is sufficient to hold the liquid in a fixed volume.

SURFACE TENSION
Molecules at the surface of a liquid are pulled inwards by the bonds. This curves the surface and gives the spherical shape of droplets.

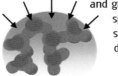

WETTING
Liquid wets a surface when the bond to the surface is appreciable compared with the bond to its neighbours.

VISCOSITY
Bond breaking creates resistance to flow. The viscosity increases with bond strength and the liquid becomes more solid-like, the structure being similar to a glass.

THIXOTROPY
Viscosity of some liquids, e.g. paints, increases with time. Agitation restores it.

MIXTURE
Liquids mix intimately if they have molecular bonds of similar type.

SOLUTION
Solids are dissolved by liquids if the bonds are similar. Ionic compounds dissolve well only in polar bonded liquids, where the charged ions are neutralised by the dipole charges.

PRESSURE
Molecular movement allows liquid pressure to equalise, though it is higher at greater depth because of the weight of liquid above. Movement of liquid under pressure is used in hydraulic power transmission.

SUSPENSION
Small solid particles in liquid.

EMULSION
A liquid may be dispersed in a second immiscible one if an emulsifying agent is added. Oil-in-water and water-in-oil are the two types. The emulsifying molecule has a water-soluble end and an oil-soluble end. Many foods, e.g. milk, and cosmetics are emulsions. Soaps and detergents are emulsifying agents. Some powders that are preferentially wetted by one of the liquids will stabilise an emulsion.

COLLOIDAL SOLUTION
If suspended particles are extremely small they do not settle under gravity and cannot be filtered out.

GEL
Colloidal particles may link together and form a solid framework that traps the liquid. Examples include gelatine, asphalt and Portland cement.

LIQUID CRYSTAL
Certain compounds flow as liquids but have an ordered long range molecular structure, as in solids. An electric field or temperature change can trigger molecular alignments and change the optical properties. Displays for televisions, laptops, monitors, calculators and watches use liquid crystals.

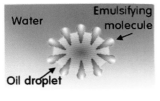

36. Liquids

37. Liquid Crystals

Some long molecules form liquids that exhibit long-range order normally found only in solids. The molecules can align in the same direction in suitable conditions, such as the application of an electric field. The liquid crystals are birefringent, meaning that an incident ray of light is split into two rays each with vibrations at right angles to those of the other. Many solid crystalline solids, minerals for example, exhibit birefringence.

Liquid crystals are used mainly in displays such as those found in televisions, monitors, laptops, mobile phones, calculators and watches. A very thin layer is sandwiched between two sheets of glass and this assembly is sandwiched between two crossed polarising sheets. Incident light is thus polarised before hitting the liquid crystal. Two rays, with vibrations at right angles emerge, and light is transmitted through the second polariser.

A transparent electrode, fixed to the glass and allocated to a pixel, carries a signal that orientates the crystals and reduces or prevents the transmission of light. Thus the illumination of the pixel is controlled.

38. Glasses

How Everything Came About

Some long organic molecules have the property of forming a liquid which flows in the normal way, but in certain conditions the structure is ordered as in a crystal. The ordering can be of alignment or alignment and position. The molecules can be rod-like or disc-shaped.

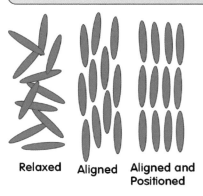

Relaxed   Aligned   Aligned and Positioned

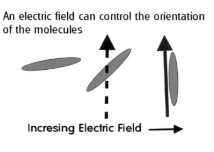

An electric field can control the orientation of the molecules

Incresing Electric Field ⟶

The crystals are birefringent and alter the polarisation of light passing through. Normally, two crossed polarisers prevent the passage of light but a liquid crystal material, placed between the two, produces two rays vibrating at right angles to each other, so light passes through the second polariser.

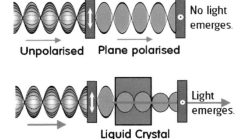

LIQUID CRYSTAL DISPLAYS

(Not to scale)

The liquid crystal layer, about 0.005mm thick, is sandwiched between two glass plates. A polymer coating on the glass is finely grooved to promote initial alignment of the molecules. Transparent electrodes, fixed to the glass, provide an electric field to rotate the molecules. Thus each pixel can be given a level of brightness.

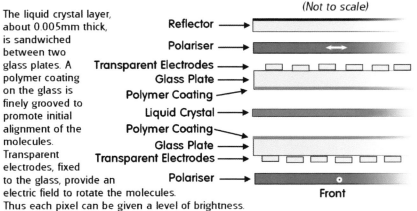

37. Liquid Crystals

38. Glasses

Some liquids, if cooled rapidly, form glasses by retaining their liquid molecular arrangement. The random arrangement of the molecules contrasts with the ordered arrangement that would be adopted under conditions of slower freezing. It is mainly molten oxides that exhibit this behaviour.

Glasses do not have precise melting points but gradually soften on heating. This is a characteristic feature of all amorphous, i.e. non-crystalline, solids. Glass is strong but brittle. The random arrangement of molecules does not provide planes along which slip can occur. Deformation is thus restricted and brittleness results. Crystalline solids differ in this respect.

Glasses are unstable and molecules eventually migrate and form crystals in the structure. Ancient glass is often no longer transparent for this reason.

Naturally occurring glasses are formed when molten rock suddenly freezes, as in lava emissions from volcanoes.

39. Polymers

Some quickly-cooled molten materials, especially oxides, form open structures with random arrangements of the molecules. Such solids (amorphous), are in effect supercooled liquids. They do not have precise melting points but gradually soften. They can flow as a liquid but very slowly, and can crystallise eventually (devitrification).

Most glasses are based on silicon dioxide (silica, SiO_2) which has a tetrahedral sructure, each silicon atom being between four oxygen atoms.

Other oxides are added to aid processing or give enhanced properties.

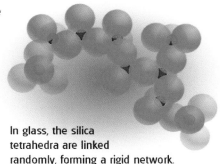

In glass, the silica tetrahedra are linked randomly, forming a rigid network.

COLOURED GLASS

Substances that absorb light of particular wavelengths are added to make coloured glass, e.g. copper for red, iron oxide for green. Reflection is from bound electrons, slightly below the surface so the colour seen is the same as the transmitted light (body colour). Crystal glass gets its brilliance from the addition of lead oxide.

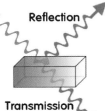

LAMINATED GLASS

A layer of plastic bonded between two sheets of plate glass prevents shattering on fracture.

TOUGHENED GLASS

The outer surface of glass is heated then rapidly cooled. This creates tension which puts the bulk of the glass into compression, hence strengthening it.

BRITTLENESS

Glass is strong but brittle because the random atomic arrangement prohibits slip planes and plasticity. Fracture is by crack growth initiated from microscopic surface cracks. Because cracks propagate under tension, glass is weaker in tension than compression.

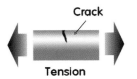

LAVA

Natural glasses are formed in lava from volcanoes by rapid cooling of molten rock.

GLASS FIBRES

Continuous fibres are woven into glass cloth. Discontinuous fibres are used for insulation and plastic reinforcement. Optical fibres are used for light transmission.

38. Glasses

39. Polymers

Polymeric materials include plastics, rubbers, resins and various natural, man-made and synthetic fibres.

A polymer is a long chain of identical molecules, usually based on carbon. The large number of possible molecules and the ability to incorporate different ones gives great potential for the development of new polymeric materials.

In polymeric materials the polymers are randomly tangled and, being flexible, render the bulk material soft. The softening increases as the temperature is increased, the material eventually flowing as a liquid.

Cross linking between the polymers modifies the properties. The material becomes stiffer or rigid. The rigid plastics do not soften at high temperature but degrade.

Elastomers have cross-linked polymers that tend to coil. They can be stretched without breaking and the large extensions possible are recovered on releasing the tension.

The wide range of fibres include natural ones from plants and animals, modified natural ones and totally synthetic ones.

As with glasses, thermal motion of the molecules can bring about a degree of undesirable crystallisation.

40. Metals and Alloys

Polymeric materials (plastics, resins, elastomers and fibres) have long chains of up to millions of molecules. They may be natural, modified (man-made) or synthetic. The molecule (monomer) is usually based on carbon and the bonding covalent. Carbon has the ability to combine with itself indefinitely.

Polyethylene has a carbon and two hydrogen atoms.

Other monomers are generally much more complex.

Polymer chains are flexible and tangled so the structure is not compact. The chains bond to each other by van der Waals forces, so the strength is limited though the chains are strong. On heating the material softens and flows (thermoplastic).

Co-polymers have lengths of a different polymer randomly located within the chain.

CRYSTALLIZATION

Crystallization of the amorphous structure can occur, totally or partially. The chains fold giving imperfect crystals (spherulites).

Random side banches inhibit crystallization. Plasticizers are added to prevent crystallization as in celluloid and cellophane. Evaporation of the plasticizer may eventually lead to embrittlement.

Side → branches

RIGID POLYMERS

As the cross-linking increases, the polymer becomes rigid. Ebonite (highly vulcanised rubber) and Bakelite are examples. Rigid plastics do not soften at high temperature but degrade (thermosetting).

ELASTOMERS

Elastomers show large reversible extension. The chains are cross-linked, bend easily and tend to coil. They are mobile at room temperature, hence low temperature causes embrittlement.
Natural rubber is a viscous liquid. In vulcanizing, sulphur atoms are added to form cross-links between carbon atoms.

Cross-links

Relaxed

Under tension

FIBRES

Polymeric natural fibres may be cellulosic (cotton, linen, etc.) or protein (wool, silk, hair, etc.). Man-made or synthetic polymers can be produced as fibres, e.g. rayon, which is regenerated cellulose, and nylon.

Carbon fibres are metallic and crystalline but are produced by charring organic polymers such as rayon. They can be very strong depending on the degree of orientation of the crystals.

39. Polymers

40. Metals and Alloys

Most elements are metals which are characterised by being good conductors of electricity and heat. Both features arise from the mobility of the outer electrons, which are not held by a particular atom but can readily move through the structure.

Metals form crystals of simple geometrical arrangements of atoms. The layers of atoms provide planes along which slip can occur. This allows deformation without fracture which is exhibited as plasticity and ductility.

As a metal solidifies from the molten state, the crystals grow and lock together as grains in a polycrystalline structure. The slower the solidification, the larger the grain size.

Alloys are mixtures of metals. Some metals mix so that the atoms of each are distributed randomly within the crystal structure. Others produce very complex molecular crystalline structures, which depend on the proportions of each metal present and on the rate at which the molten mixture is cooled. The structures and resulting properties can be further modified after solidification by a further heating cycle or mechanical deformation.

41. Crystalline Solids

About three quarters of all elements are metals. The orbits of electrons in unfilled shells overlap in neighbouring atoms.

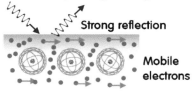

Strong reflection

Mobile electrons

This provides the metallic bond and allows mobility of electrons through the lattice. Hence metals conduct electricity and heat readily.

Light photons cause currents in the mobile electrons, giving strong reflection. The colour seen is surface (not body) colour, the (albeit weak) transmitted light being of the complementary colour.

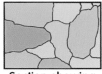

Yellow reflected light

Thin gold foil

Blue-green transmitted light

SINGLE CRYSTAL

Because metallic bonding is not directional, the atoms form crystals with simple arrangements.

POLYCRYSTALLINE STRUCTURE

As molten metals solidify, crystals grow and join together. Each crystal is a grain. Fast cooling gives small grain size while slow cooling gives large grain size.

Section showing grain structure

ELASTICIY

Extension or contraction from an applied force is resisted by the atomic bonds. The shape recovers when the force is removed.

PLASTICITY

If the force is excessive, the atoms can slip into new positions. The deformation is permanent. Plasticity is the reason for ductility in metals.

FRACTURE

Eventually the force overcomes the strength of the atomic bonding.

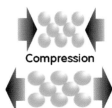

Compression

Tension

slip

FATIGUE

Repeated deformation can produce fracture by growth of cracks, often from internal or surface defects.

ALLOYS

Alloys are mixtures of different metals. Small amounts of different metals are frequently added to metals to increase their hardness and strength.

SOLID SOLUTIONS

Some metals are soluble in each other and when solidified form crystals with different atoms distributed randomly. Examples are copper-nickel, iron-nickel, gold-silver, gold-copper.

MULTIPHASE ALLOYS

Depending on the number and relative proportions of the metals, and the rate of cooling, the alloys consist of grains of various phases (metals, solid solutions or compounds having fixed chemical compositions). Examples are iron-carbon (steel), copper-zinc (brass), copper-tin (bronze), lead-tin (solder).

40. Metals and Alloys

41. Crystalline Solids

Most non-metallic crystalline solids are compounds and most are minerals.

The atoms making up the molecule differ in size and bonding requirements and are forced into particular arrangements. They form a three-dimensional lattice with a repeating structure. The lattice geometry is responsible for the form adopted by the crystal as it grows.

The crystal grows from a liquid which may be a molten state. Slow cooling, which gives time for the atoms to diffuse to their preferred positions, gives large crystals. The liquid may be a solution and this also can result in growth of large crystals.

Minerals solidifying from molten rock have their natural crystal form suppressed by the need to lock together in a polycrystalline mass.

Gems, which may be natural or synthetic, are minerals that have durability and a particularly attractive appearance.

Ceramics are made from clay which consists of minerals. The firing of the ceramic forms a glass matrix so the finished material could be classed as a composite.

42. Composite Materials

How Everything Came About

A few elements are non-metallic solids at normal temperatures and some of these are crystalline. However, most non-metallic crystalline solids are compounds.

CRYSTAL FORMS
The complex, but symmetrical, crystal forms are determined by the atomic arrangement.

HARDNESS
Covalent bonding and distinguishable individual molecules gives soft low-melting-point solids because the bonding between molecules is weaker than within molecules. Continuous structures (ionic and some covalent) give the reverse.

MINERALS
Minerals are naturally occurring non-metallic crystalline solids and are the main constituents of rocks. Most metals are found in nature not pure but as minerals in combinatiion with non-metals.

GEMS
A gem is a mineral having beauty and durability. Corundum (alumina, aluminium oxide, Al_2O_3), e.g., is used as an abrasive (emery) and in grinding wheels, but is also sapphire or ruby depending on the colour.

CORUNDUM

CERAMICS
Clay is a fine-grained material, mainly silica, alumina and water. Its plasticity allows it to be shaped, and then fired to give a hard solid consisting of a glass binding mineral crystals together. Bricks, porcelain, electrical insulators and magnets are thus produced. Refractories are high melting point solids, of similar composition, used for heat insulation at high temperatures, e.g. furnace linings.

N.B. 'ceramic' is often taken to include minerals and glasses.

The atomic arrangement is determined by
1. Relative sizes of the atoms,
2. Overall electrical neutrality,
3. Bond (usually covalent) requirements, in number and directionality,
4. Minimum repulsion between nuclei,
5. Minimum distances between atoms

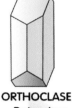

QUARTZ
Silica (silicon dioxide, SiO_2)

ORTHOCLASE
Potassium aluminium silicate ($KAlSi_3O_8$)

MICA
Potassium aluminium silicates with other elements (sodium, lithium, magnesium)

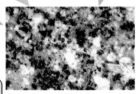

GRANITE
Mineral crystals grow and interlock, as the molten rock solidifies.

MOHS HARDNESS SCALE
Soft
1. Talc
2. Gypsum
3. Calcite
4. Fluorite
5. Apatite
6. Orthoclase
7. Quartz
8. Topaz
9. Corundum
10. Diamond
Hard

FRACTURE
Fracture is brittle because the complex atomic arrangements inhibit slip plane development. Cleavage is a clean splitting parallel to a crystal face, as commonly observed with mica.

41. Crystalline Solids

42. Composite Materials

Numerous materials have been manufactured by combining existing materials of different kinds. The drive to develop such materials is from the need to enhance or combine properties. In many single-type materials certain properties are mutually exclusive.

Composites can be classed as reinforced or bonded.

In reinforced materials, a bulk material is given, for example, increased strength, by adding fibres or particles of a different material.

In bonded materials, fragments of material with desirable properties, hardness for example, can be cemented together to give a material that can be shaped for particular applications.

Many natural biological materials, such as bone, are composites and it is by studying these that ideas for manufactured composite materials have often arisen.

43. Nanotechnology

How Everything Came About

Two or more materials can be combined to give a material with enhanced properties. Usually, one material, the matrix, is continuous and contains filaments or particles of other materials. In reinforced materials the matrix predominates, in bonded materials the filaments or particles predominate. Composites are constantly being developed.

FILAMENT: Glass, Graphite, Boron, Aramid (Kevlar), Metal, Whisker (single crystal filament)

PARTICLE: Ceramic, Metal

POLYMER MATRIX

Discontinuous fibres

Continuous fibres

Fibreglass (glass fibres in a resin matrix)

Filaments or particles are added to increase strength, toughness, stiffness, or dimensional stability, or to decrease cost. Numerous applications include parts for aircraft, cars, boats, rockets, and sporting equipment such as tennis rackets, golf clubs and fishing rods. Medical applications include lightweight aids and implants. Electrically conducting plastics are also produced.

CERAMIC MATRIX

A ceramic matrix is less common. Concrete and ferrites are important examples.

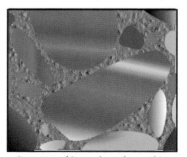

Concrete (Gravel and sand in a matrix of Portland cement.)

METAL MATRIX

A metal matrix allows use at high temperature. Filaments or particles are added to a weaker but less brittle matrix to give a strong stiff tough material for structural applications. Unique electrical and magnetic properties can also be achieved. Cermets have ceramic particles imparting strength, hardness, and resistance to wear and corrosion. They are used for cutting tools, anti-wear parts and engine and turbine components.

Cemented Tungsten Carbide (A cermet having tungsten carbide crystals in a cobalt matrix.)

The definition of a composite material is not exact. For example, steel-reinforced concrete would hardly be included but plywood might. Structures similar to composites can be formed in some alloys.

NATURAL MATERIALS

Many biological materials such as bone, bamboo and muscle, are composite and often provide ideas for synthetic materials.

42. Composite materials

43. Nanotechnology

Nanotechnology is a relatively recent technology dealing with things having size measured in nanometres. A nanometre is a millionth of a millimetre, which puts the area of interest between molecular sizes and the limits of optical microscopy.

Miniaturisation has proceeded rapidly in electronics. The ability to produce small integrated circuits each containing billions of transistors demonstrates the current capability. Extension of the established techniques of etching and micro-machining constitutes the top-down approach to engineering in the scale of nanometres.

The bottom-up approach uses deposition techniques and the manipulation of individual atoms and molecules to assemble new structures.

The technology is rapidly growing. The perceived benefits, particularly in medical diagnosis and treatment, and in information and communication applications, are very great.

Unwarranted scare stories have arisen in connection with nanotechnology. The uncontrolled spreading of hordes of self-replicating molecular robots that would obliterate civilisation has been suggested.

How Everything Came About

> Nanotechnology is involved with materials and fabricated items of sizes between about 0.1 and 100nm, one nanometre (nm) being one millionth of a millimetre. For comparison, a water molecule is 0.5nm and viruses range from 10nm to 300nm.

As particles are reduced in size their properties become quite different to their bulk properties, because, firstly, there are more surface atoms compared with interior atoms, and, secondly, quantum mechanical effects become important.

TOP-DOWN TECHNOLOGY
Machining and etching techniques, as used, e.g. in integrated circuit production, are used to form tiny features from bulk material.

Quantum dot
Particles of semiconductor material, re-emit light of a colour dependant on their material shape and size.

BOTTOM-UP TECHNOLOGY

SELF-ASSEMBLY
Deposition techniques can take advantage of the tendency of some materials to organise themselves in ordered arrays

MANIPULATION
Individual atoms can be moved in order to build molecular structures.

Optical tweezers
Transparent molecules refract more light in the centre of the beam than at the edge. Reaction pushes the molecule up and to the centre.

Graphene
Sheets of carbon, one atom thick, are stronger than steel, better conductors than copper, light, flexible, tough and transparent. An enormous number of potential applications are being developed. Some other materials have been produced as one-atom-thick sheets. Carbon nanotubes are produced as single wall and multiwall. They have similar remarkable properties and can be used to store hydrogen for fuel cells.

Atomic Force Microscope

The instrumented microscope probe can detach and move single atoms.

APPLICATIONS

> The potetial for nanotechnology is great but many applications are at a very early stage.

ENERGY
Cheap efficient solar cells
Fuel cells   Lighting

MEDICAL
Tracking particles for diagnosis
Internal body sensors
Targetted drug delivery
Timed drug release
Gene analysis and sequencing
Manipulation of body cells
Stronger, lighter implants
Biodegradable implants
Improved biocompatibility of implants

INFORMATION
Data storage devices
Optical display devices
Optical communication
Optical computing
Quantum computing
Molecular electronics
Organic electronics
Security tags

MECHANICAL
Stronger materials
Lubricants
Wear-resistant materials
Reinforced composites
Tethered satellites
Displacement sensors
Force sensors
Molecular motors

33. Molecules

44. Chemical Processes

Molecules assemble or split in chemical reactions. Energy (usually heat) may be produced, leaving the chemical products in a more stable state, or energy may have to be supplied, leaving the products in a less stable state.

Molecules thus store energy which can be released or transferred in chemical processes.

Natural chemical processes occur on Earth because of the presence of the atmosphere and water.

Oxygen readily combines with other atoms or molecules to produce oxides, energy being liberated. Rocks are largely based on silicon oxide. Metals, except gold, are found in the form of oxides.

Animal metabolism is powered by the energy produced from oxidation of food, and organic matter is decayed by oxidation via bacteria.

The atmospheric supply of oxygen is maintained by plant life. Photosynthesis allows plants to absorb energy from sunlight and liberate oxygen.

A great variety of chemical processes is found in the chemical industries of the world. There are well over 10 million unique chemical compounds that have been separated or synthesised. Chemical engineering is involved in the production of almost everything we use, yet is one of the more recent branches of engineering.

45. Life

How Everything Came About

By breaking and/or making bonds between atoms in chemical reactions, compounds are formed. Some reactions require energy (endothermic), some release energy (exothermic). Reactions generally speed up at higher temperatures. Several million compounds are known, most of them organic (involving carbon) because carbon can unite with itself indefinitely and produce large molecules.

Constituents of the atmosphere, surface water and dissolved solids are in intimate contact with Earth's surface.

 Oxygen O_2

Carbon dioxide CO_2

 Water H_2O Nitrogen N_2

 Ozone O_3 produced from O_2 by sunlight

 Gases from volcanoes

OXIDATION

Most elements form oxides in the presence of oxygen, with the release of heat (usually), light or electricity. Metals occur in nature as oxides. Gold is exceptional.

CORROSION

Oxygen, water and impurities corrode metals and alloys. Rust, for example, is iron oxide.

WEATHERING

Oxygen, carbon dioxide, water and plants break down rocks and structures, depositing products in oceans, clays and soils.

COMBUSTION

Burning is slow oxidation of a fuel: explosion is rapid oxidation. Spontaneous combustion may start from heat from micro-organisms, or in powders which provide a large surface area for oxidation.

RESPIRATION

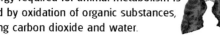

The energy required for animal metabolism is provided by oxidation of organic substances, producing carbon dioxide and water.

PHOTOSYNTHESIS

In chlorophyll-containing plants, carbon dioxide, water and sunlight produce carbohydrates, e.g. glucose. Proteins are then synthesised giving plant tissue and, via the food chain, animal tissue.

CATALYSTS

Some elements and compounds can trigger a reaction without being consumed in the reaction. Catalytic converters in automobile exhausts use metals such as platinum to remove pollutants.

INDUSTRIAL PROCESSES

Almost all man-made substances rely on chemical processes which involve the rearranging of bonds between atoms to produce molecules.

DECAY

Bacteria metabolism decomposes organic matter by oxidation. Sugary substances give useful products (fermentation). Proteins putrify.

44. Chemical Processes

45. Life

It is not easy to stipulate exactly the difference between living and non-living things. Perhaps the principal characteristic of life is the ability to reproduce, though not all animals can reproduce.
The various kinds of life forms, namely, animals, plants, fungi, bacteria and viruses, have very different appearances but operate using very similar processes. This is indicative of their common origin.

All are made mainly of proteins which are large polymers. The number of different proteins involved is very large, yet all are made from arrangements of only about 20 amino acids. The amino acids are a group of related carbon-based molecules.

Plants and some bacteria take in energy from sunlight, absorb carbon dioxide from the atmosphere and release oxygen. Animals, fungi, and other bacteria absorb oxygen and obtain energy by oxidising food, which consists of tissue from animals or plants.

All life forms are made up of cells. Each cell is supplied with a source of energy and has its own processing units to maintain and reproduce itself, and to manufacture proteins for specific applications in the living individual. Each cell carries its instructions in the form of a copy of the DNA molecules specific to the individual.

46. Origin of life   48. Bacteria and Viruses   49. Plants
50. Animals

How Everything Came About

> The most outstanding characteristic of living organisms is their ability to reproduce. The offspring are similar to the parent but nevertheless show small differences which, over long time periods, give rise to appreciable changes in populations.

The varieties of life forms, represented by animals, plants, fungi, bacteria and viruses, have widely different external appearances but very similar structure and mechanisms.

PROTEINS

All living matter is made of large molecules (polymers) called proteins, which are made from about 20 amino acids. Chains of hudreds or thousands of amino acids, each having a different arrangement, form different proteins, The number of possible different proteins is very large.

DNA

Organisms preserve their individuality and transmit it to their offspring by DNA (deoxyribonucleic acid) or RNA (ribonucleic acid). DNA is a molecule having a long series of combinations of 4 substances. The number of possible different DNA molecules is very large.

AUTOTROPHS
(Plants, some bacteria)

Carbon dioxide and water are broken down in the chloroplast by solar energy to give glucose and oxygen, oxygen being released.

HETEROTROPHS
(Animals, fungi, bacteria)

Food (organic compounds) and oxygen are broken down in the digestive system to give glucose, water and carbon dioxide, carbon dioxide being released.

Membranes provide large surface areas for the chemical reactions

CHLOROPLAST

MITOCHONDRION

Cytoplasm

Sap

Nucleus with DNA

Ribosome

PLANT CELL

ANIMAL CELL

CELL RESPIRATION

Glucose is used by plant and animal cells to produce adenosine triphosphate (ATP) from adenosine diphosphate (ADP) and phosphate. ATP is the energy supply for the cell's activities, such as protein synthesis, transport of products or muscle contraction. ATP reverts to ADP and phosphate as the energy is used. The production sites are the mitochondria.

DNA

Ribosome

BACTERIUM
(UNICELLULAR)

Plant and animal cells vary according to location and function. Bacteria have numerous varieties. Typical cells with some of their main constituents are shown (not to scale).

45. Life

46. Origin of Life

It is not known how life started on Earth, but it is known that it started early, around the time that the temperature reduced below the boiling point of water and the intense bombardment by asteroids ceased.

Although life started soon, it was only relatively recently, within the last 600 million years, that complex life evolved.

This suggests that the step from small simple lifeforms to larger complex forms is more difficult than the evolution of simple microscopic forms.

However, it has not even been possible to produce the simplest of life forms in the laboratory. Proteins can be synthesised but the required steps beyond this have not been achieved.

It has been suggested, with some justification that life arrived on Earth on meteorites. This is possible but does not of course, explain the origin of life.

47. Extraterrestrial Life

How Everything Came About

It is not known how life started, though present knowledge can provide hypotheses.

WHEN?
Life originated on Earth about 4 billion years ago, which is relatively soon after the origin of Earth. Life started when the surface temperature dropped below the boiling point of water and asteroid bombardment ceased.

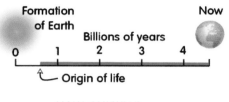

HOW QUICKLY?
Possibly in less than a million years, which gives little time for low-probability random events.

WHERE?
Few if any land masses existed and Earth's surface was subjected to deadly solar ultraviolet radiation. Life therefore probably started in the ocean. Hydrothermal vents, the present home of extremophiles, appear a likely location. Alternatively, it is possible that simple life forms arrived from space on meteorites.

HOW?

SMALL ORGANIC MOLECULES
Amino acids

These can be easily produced from chemical mixtures and ultraviolet energy input.

Many organic molecules have been detected in space, including glycine, a key biochemical amino acid. Meteorites have yielded similar finds.

JOINING TO FORM LARGE MOLECULES (POLYMERS)
Proteins

These also can be synthesised but their natural origin is more difficult to explain.

Glycine (NH_2CH_2COOH)

Nucleic acids

HEREDITY

RNA

DNA

This is a very difficult step which cannot be reproduced in the laboratory.

DNA
There is only one genetic code on Earth, common to all life forms.

DEFINITION OF LIFE
There is no satisfactory definition of life. Features of life include growth, reproduction, metabolism and evolution. A virus has no independent life when isolated but seems alive when in combination with a host.

46. Origin of Life

47. Extraterrestrial Life

There is no evidence of intelligent life anywhere else in the Universe. Listening for meaningful signals continues within the SETI project, and there have been signals transmitted from Earth into space. Nothing positive has been found.

Signals cannot travel faster than the speed of light. This implies extremely long journey times for signals across even a tiny fraction of the Universe. So the absence of meaningful signals does not in itself rule against there being life elsewhere.

The Drake equation was introduced to quantify the likely number of communicating extraterrestrial intelligences in our galaxy. However, it cannot resolve the issue, because the predicted number depends on whether the estimates of the factors in the equation are chosen optimistically or pessimistically.

In spite of the likely large number of planets in the Universe, there are good grounds for arguing that the Earth is special. There are many coincidental factors that produce the particular conditions suitable for the complex life we see here.

Exobiology and astrobiology are areas of study concerned with possible life forms that could originate, exist and evolve in environments beyond Earth.

How Everything Came About

There is no reliable evidence for extra-terrestrial intelligent life anywhere in the universe. Though the search will continue, the likely situation is that simple life may be fairly common while complex intelligent life is rare or even unique to Earth.

COMMUNICATION WITH ALIENS

LISTENING

The search for extra-terrestrial intelligence (SETI) has involved listening for meaningful radio signals from space since the 1960s. More recently, searching for visible (laser) signals has been included. Nothing definite has been detected.

TRANSMITTING

Very few signals have been transmitted. Signals sent in all directions require high power and would be prohibitively costly. Narrowly beamed signals can be sent over great distances but how are the targets to be chosen? Earth's own radio, TV, etc., communications have been spreading through space, albeit weakly. By now the earliest will have reached only a small fraction of the way across our own galaxy.

DRAKE EQUATION

The number of communicating extraterrestrial intelligences (ETI) in the galaxy (N) can be calculated by the Drake Equation. However, the result can be very large or almost zero depending on whether high or low values are assumed for the contituent numbers in the equation.

DRAKE EQUATION
$$N = R \times f_p \times n_e \times f_l \times f_i \times f_c \times L$$
R = yearly rate at which stars form
f_p = fraction of stars having planets
n_e = number of planets per star with environments suitable for life
f_l = fraction on which life actually develops
f_i = fraction which develop intelligent life
f_c = fraction capable of communicating
L = time in years devoted to communicating

IS THE EARTH SPECIAL?

Some of the many factors that make Earth particularly suitable for life:

1. Right position in right kind of galaxy: enough heavy elements.
2. Right distance from star: liquid water.
3. Right mass of star: long enough lifetime, and limited ultraviolet radiation.
4. Large neighbouring planet (Jupiter): protects Earth from comets and asteroids.
5. Large moon: stabilises Earth's tilt.
6. Right planet mass: retains atmosphere and oceans, and allows plate tectonics.
7. Right tilt: moderate seasonal changes.
8. Right ocean-land split: evolution.
9. Right amount of carbon: complex organic molecules, and limited greenhouse effect.

It may be that we are alone and will eventually colonise the universe. At present, distances are too great and life too short to contemplate such journeys. However, genetic engineering and medical advances may extend human life enormously, even indefinitely, and make such journeys feasible.

45. Life

48. Bacteria and Viruses

Bacteria are single-celled organisms and are the closest relatives of the original life forms on Earth. They can live in a wide range of environments and because of their short life times they can evolve quickly to adapt to changes. Although they usually reproduce by the cell dividing into two, some mating occurs which gives rise to mutations.

Many bacteria are useful, indeed some are essential in the digestive systems of animals. They are responsible for many diseases but compensate to some extent by providing antibiotics.

Viruses are the smallest life form, being no more than about a tenth the size of a bacterium. They are a borderline life form as they have no independent life, but live only as parasites in the cells of bacteria, plants and animals.

They compete with the genetic instructions of the host cell for control of its activities and reproduce by manufacturing parts in the cell. Mutations arise in the assembly.

Like bacteria, viruses are responsible for many diseases.

How Everything Came About

BACTERIA
As bacteria can survive in almost any environment, their metabolism varies widely. They are unicellular, varying in type, shape and size. Some can move using hair-like filaments, others can creep or glide. The cell contains a single circular strand of DNA.

FERMENTATION
Some anaerobic bacteria produce partially oxidised substances. The processes give relatively little ATP for energy so large quantities of fermentation products result. An important process in the food industry is the conversion of sugar to lactic acid by the bacteria Lactobacillus. Yeast, however, used extensively for fermentation, is a fungus.

REPRODUCTION
Colonies can double in size in minutes. Usually the cell divides into two, but some mating occurs, one cell passing some DNA to another. This gives rise to mutations and is a process utilised in genetic engineering.

SPORES
Some bacteria, when nutrients become restricted, form dormant spores which are resistant to radiation, desiccation, heat and chemicals. The spore can return to a metabolically active state.

EXTREMOPHILES
Around deep-ocean volcanic rifts, bacteria and similar microbes live in temperatures as high as 120°C and pressures up to 400 atmospheres. Other extremophiles have been found living in icebergs and 3.5km below the earth's surface.

DISEASE
Although bacteria are useful, for example in the digestive systems of animals and by decomposing dead tissue, they cause disease. They attach to a host cell and may destroy it. Alternatively, the host may suffer irritation from bacteria waste products, from bacteria toxins or from exaggerated response of the host's immune system.

**Salmonellosis Cholera
Tuberculosis Anthrax
Scarlet fever Typhoid
Legionnaire's disease
Meningitis Tetanus
Diphtheria Leprosy
Whooping cough**

VIRUSES
Viruses, the smallest life form (about 0.0001mm), vary in type and shape, and consist essentially of a length of DNA or RNA within a protein coat. They are parasitic, infecting bacteria, plants and animals, and have no independent life.

REPRODUCTION
Viruses reproduce by making virus parts and assembling them within the host cell. Mutations arise in the assembly.

DISEASE
A virus penetrates the host cell and competes with the genetic material for control of cell processes.

**Shingles Rabies Chickenpox
Meningitis Polio Mumps CJD
German measles Influenza
Measles Glandular fever
Yellow fever HIV**

45. Life

49. Plants

Plants obtain energy from sunlight. The photons cause chlorophyll molecules to release electrons which allow ATP (adenosine triphosphate) to be produced from phosphate and ADP (adenosine diphosphate). ATP is an energy source vital in the metabolism of plants and animals.

Water is absorbed, the hydrogen retained and oxygen released. Carbon dioxide is absorbed and with the hydrogen and the energy from the ATP is processed to glucose. The ATP reverts to phosphate and ADP.

The glucose is processed by the cells, in a similar manner to that employed by animal cells, to yield further ATP. Surplus glucose is stored as starch.

Energy is used in plant movement, transport of constituents and building of proteins.

Two sets of tubes provide the means of transporting substances around the plant. One distributes the water and the second set carries the products of photosynthesis.

Growth of plants is by cell growth and cell division. Cell division is also a means of reproduction in which the offspring are genetically identical to the parent. Some bisexual reproduction occurs, as in animals, and allows mutations to arise.

Plants are multicellular, each cell having an enclosed nucleus which contains chromosomes. The cell differs from an animal cell in having a firm wall of cellulose.

PHOTOSYNTHESIS

Plants (excluding parasites) get energy from sunlight. Photons excite and release electrons in chlorophyll molecules. The electrons permit the synthesis of ATP (adenosine triphosphate) from ADP (adenosine diphosphate) and phosphate, and, with hydrogen from water, the synthesis of NADPH from NADP (nicotinamide adenine dinucleotide phosphate).

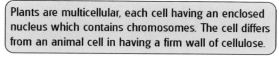

Enzymes control the synthesis of glucose from carbon dioxide using the hydrogen from NADPH and energy from ATP. The glucose is stored as starch or fed to cells for production of amino acids or lipids, via glycolysis and the TCA (tricarboxylic acid) cycle.

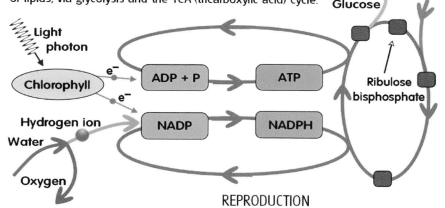

Plants are green because chlorophyll absorbs light strongly in the red and blue ends of the spectrum but less so in the central green band.

GROWTH

Growth is by cell division (mitosis) and cell growth. Plants, unlike animals, have almost unlimited growth potential.

REPRODUCTION

Asexual reproduction by cell division is manifest in spores, shoots, runners, tubers and bulbs. Offspring are clones, being genetically identical to the parent.

TRANSPORT

Dead tubes (xylem) take water up the plant. Living tubes (phloem) take the products of photosynthesis around the plant.

Bisexual reproduction, as in animals, occurs by meiosis. The male (pollen) and female (ovule) gametes combine to produce a seed surrounded by fruit.

FUNGI
Fungi are no longer classed as plants. They feed (like animals) by digesting food from other organisms. They play a vital role in providing food and drugs, and decomposing, but also cause disease and destroy crops.

45. Life

50. Animals

Animals vary greatly in size and appearance but employ similar processes in their functioning.

All have muscle constituting body mass and most have a skeleton, internal or external, consisting of rigid material.

The two control systems, electrical and chemical, are common to all animals, but the electrical nervous system ranges from simple nerve cells in the lower animals to complex networks and a brain in the higher forms.

The senses, activating the nervous system, are similar throughout, but are more or less developed in different types of animal.

All take in oxygen though the methods differ.

Food is digested by organs in all but the lower animals, the latter digesting in individual cells.

Most have a circulation system to supply the cells with their needs and remove waste products but some of the lower animals use their water environment to provide transport to and from the cells.

Reproduction is based on the division and combination of cells, each new cell carrying copies of genes from the parent cells.

51. The Human Frame

How Everything Came About

> Though varying greatly in appearance, animals are similar in the organisation of their bodies. Evolution has increased the complexity of activity from cell-based to tissue-based and finally organ-based.

HUMANS AND OTHER VERTEBRATES

FRAME
Internal jointed skeleton of bone
Muscle
Skin

CONTROL
Brain
Nervous System (electrical)
Endocrine System (chemical)

CIRCULATION
Heart
Blood

RESPIRATION
Lungs

DIGESTION
Small Intestine

Digestion, respiration and circulation are intimately linked in converting food to simpler chemicals, distributing them to all parts of the body and removing unwanted products.

REPRODUCTION
Genes

Arrows in/out: SENSES, OXYGEN, CARBON DIOXIDE, FOOD, WASTE, MOVEMENT, HEAT

INVERTEBRATES

Arthropods have a segmented body, jointed limbs and an external chitinous skeleton. Echinoderms have an internal calcareous skeleton. Molluscs and brachiopods have an external calcareous shell. Annelids, coelenterates and sponges have soft bodies though some coelenterates have a supporting calcareous base (coral) and sponges have supporting mineral fibres. All have muscle.

All have nerve cells which form a network in all but sponges. A dense central region of the network forms a brain in some invertebrates. The range of senses is similar throughout, but particular senses may be absent, weak or strong. All produce hormones for chemical control.

Terrestrial invertebrates respire through the skin or via tubes feeding directly to the cells. Aquatic invertebrates use gills or respire via individual cells.

Digestive organs are present in all but coelenterates and sponges which digest via individual cells.

Arthropods, molluscs, echinoderms and brachiopods have an open circulation system, in that the heart pumps blood into a body cavity but does not return it. Coelenterates and sponges have no heart (nor any organs). The water environment is used for the circulation system.

Reproduction is via the copying of genes from cell to cell and parent to offspring.

50. Animals

51. The Human Frame

The human frame consists of a jointed skeleton of bone, surrounded and held in position by muscle which constitutes the body's flesh. The flesh is covered by a layer of skin.

Ligaments hold the bones of the skeleton together at the joints. Muscles provide the forces to move the bones and internal organs.

Muscle action relies on the interaction of two proteins, actin and myosin. A nerve impulse starts a sequence of events leading to a rapid contraction and relaxation of the myosin. This acts in a ratchet fashion against the actin to contract the muscle.

The skin carries nerve fibres to transmit sensations to the brain, and sweat glands to give temperature regulation.

Beneath the skin is a layer of fat, which provides insulation from heat loss and serves as a store of surplus energy that the body can draw on if necessary.

52. Body Circulation System

How Everything Came About

> A skeleton of bones is connected by muscles and enclosed in an outer layer of skin.

BONE TISSUE

A central channel circulates cytoplasm to cylindrical layers of cells and contains a blood vessel supplying blood via capillaries.
A hard matrix of collagen (a tough protein) fibres in calcium salts surrounds the cells.

Spongy Bone — Porous structure.

Compact Bone — Dense and strong.

Ligament — Tough bands of collagen holding the joint in place.

Cartilage — The cells are in a matrix of collagen fibres and a firm gel. There are no blood vessels.

Marrow — Blood cells are manufactured in the marrow.

Tendon ← **Muscle**
Collagen attatching muscle to bone.

MUSCLE

Movement of bones and organ tissue is by contraction of muscles, which form the body's 'flesh'. Animal muscle is lean meat.

MUSCLE ACTION

Muscle fibre contains a bundle of fibrils. Each consists of interleaved filaments of the proteins actin and myosin.

Nerve Fibre, **Cisterna**, **Fibril**

In a relaxed muscle, bonding sites on the actin are blocked.
A nerve impulse releases calcium ions from the cisternae, unblocking the sites. Myosin-actin bridges form. ATP attaches causing contraction of the myosin. The actin moves along and the myosin recovers. The process repeats rapidly in a ratchet fashion. When the stimulus stops, calcium ions are pumped back, blocking the sites again.

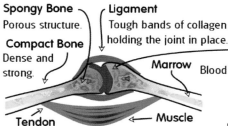

Actin, Myosin, Ca^{++}, ATP

SKIN

Keratin, a strong fibrous protein, forms the dead outer layer and also hair and nails.

Hair, **Blood Capillaries**

Epidermis — Cells fill with keratin and move outwards to renew outer layer.

Dermis — Collagen and other fibres.

Sweat Gland

Layer of Fat — Oil occupies most of each cell.

Nerve Fibre

> **TEMPERATURE REGULATION**
> If the body is too hot, blood vessels open to pass more heat to the skin, and glands produce more sweat to remove heat by evaporation.

How Everything Came About

51. The Human Frame

52. Body Circulation System

Food is transported via the stomach to the small intestine where most of the processes of digestion take place. Undigested material is disposed of via the large intestine.

    Blood is circulated to all parts of the body, every living cell depending on the supply of nutrients and removal of waste. Oxygen is carried from the lungs and carbon dioxide returned. Digested food products are taken from the small intestine to the liver and from there to parts of the body where they are required. Surplus water is collected from the large intestine and taken to the kidneys for disposal.

    The blood also carries hormones from the endocrine glands to various parts of the body.

    Heat is produced by the oxidation in the digestive processes and carried by the blood to all parts of the body.

    The lymphatic system produces antibodies and distributes them via lymph tubes and the blood stream.

53. The Immune System   54. Food

How Everything Came About

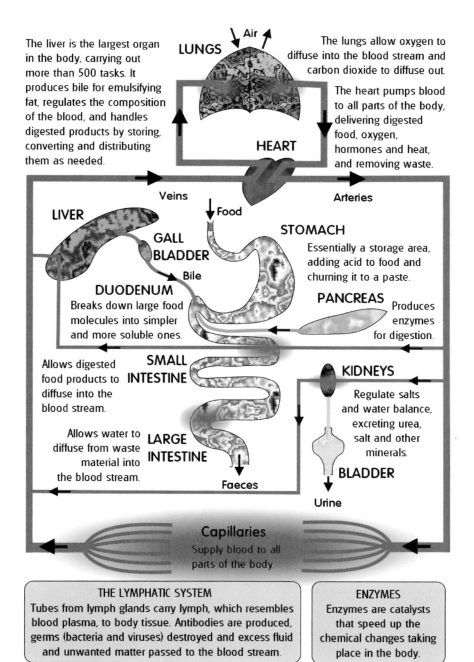

The liver is the largest organ in the body, carrying out more than 500 tasks. It produces bile for emulsifying fat, regulates the composition of the blood, and handles digested products by storing, converting and distributing them as needed.

LUNGS — Air
The lungs allow oxygen to diffuse into the blood stream and carbon dioxide to diffuse out.

HEART
The heart pumps blood to all parts of the body, delivering digested food, oxygen, hormones and heat, and removing waste.

Veins — Food — Arteries

LIVER

GALL BLADDER — Bile

STOMACH
Essentially a storage area, adding acid to food and churning it to a paste.

DUODENUM
Breaks down large food molecules into simpler and more soluble ones.

PANCREAS
Produces enzymes for digestion.

SMALL INTESTINE
Allows digested food products to diffuse into the blood stream.

KIDNEYS
Regulate salts and water balance, excreting urea, salt and other minerals.

LARGE INTESTINE
Allows water to diffuse from waste material into the blood stream.

BLADDER

Faeces — Urine

Capillaries
Supply blood to all parts of the body.

THE LYMPHATIC SYSTEM
Tubes from lymph glands carry lymph, which resembles blood plasma, to body tissue. Antibodies are produced, germs (bacteria and viruses) destroyed and excess fluid and unwanted matter passed to the blood stream.

ENZYMES
Enzymes are catalysts that speed up the chemical changes taking place in the body.

52. Body Circulation System

53. The Immune System

An antigen is a substance, usually some form of protein, that the body recognises as an infection.

White blood cells are of several types and have the specific task of attacking antigens which they do in various ways. The white blood cells are manufactured in the lymph nodes and in bone marrow.

Some white cells produce antibodies which can, after doing their work, remain in the blood stream and provide long-term protection against specific diseases. In the absence of naturally produced antibodies, vaccines can be administered to trigger the antibody production. Vaccines are dead or harmless germs. A more rapid protection can be given by introducing antibodies from previously infected people or animals.

A proportion of people have red blood cells that carry antigens, and all of us have antibodies that will attack antigens of types not present in our own blood. Because of this, it is necessary to define blood groups according to the presence or absence of particular antigens. Knowledge of a person's blood group ensures that any donated blood is compatible.

How Everything Came About

The body's immune system is triggered by antigens. These are proteins or complex carbohydrates and may be constituents of micro-organisms or transplanted tissue.

Skin contact　　Wounds　　Vaccination　　Inhalation　　Ingestion

ANTIGENS

FIRST LINE OF DEFENCE　　Skin　　Blood clots　　Tears　　Saliva　　Mucus　　Stomach acid

WHITE BLOOD CELLS

White blood cells are made in the lymph nodes and bone marrow and are of several types. They circulate in the blood and lymphatic system and attack antigens in various ways. Some can change shape and 'crawl' between tissue cells.

NON-SPECIFIC IMMUNITY

Monocytes engulf and digest antigens.

Granulocytes release destroying substances.

CELLULAR IMMUNITY

T-Lymphocytes bind to infected, cancer or transplanted cells and kill them.

HUMORAL IMMUNITY

B-Lymphocytes bind to specific antigens and produce antibodies. The binding sites match in size, shape and charge distribution on the molecules. The antibodies bind to and neutralise or destroy the antigens and remain in circulation, giving long-term protection against specific diseases.

PASSIVE IMMUNITY

Antibodies from previously infected people or animals are administered to give immediate immunisation against fast-acting toxins or microbes.

RED BLOOD CELLS

Red blood cells carry antigens. The body produces antibodies naturally, or in response to the antigens in transplanted blood. There are about 200 blood groups, in 19 known systems, but most do not form antibodies sufficient to cause problems in transfusion.

IMMUNISATION
Antibody production can be triggered by introducing vaccines, which are dead or harmless forms of germs.

GROUP A　　GROUP B　　GROUP AB　　GROUP O

Antibody　　Antigen

The ABO system is of prime importance in blood transfusion because the antibodies are always present when the corresponding antigen is lacking. Transfused cells with foreign antigens would be attacked by host antibodies. One of the Rhesus systems has also to be taken into account, blood being designated Rh+ or Rh-.

52. Body Circulation System

54. Food

Food is required by the body to provide energy, raw materials for its structure and substances for the regulation of its processes.

Energy comes from the oxidation of organic compounds in the digestive system. These compounds are classified as sugars, fats or proteins. Fats provide about twice the energy of sugars or proteins, weight for weight.

The body is assembled from proteins and the building blocks for these are the amino acids. About half of the amino acids needed can be synthesised within the body from food products. The others must be obtained directly from the proteins in food. Water, and a number of minerals are also required for assembling the structure of the body.

Regulation of the body's processes requires water, fibre and small quantities of vitamins and certain minerals. None of these is digested and thus do not contribute to the body's energy supply.

Manufactured foods contain additives. Some are traditional preservatives such as salt or vinegar, but most are recently employed food enhancers. Many of the additives are identified by E-numbers.

55. Human Metabolism

How Everything Came About

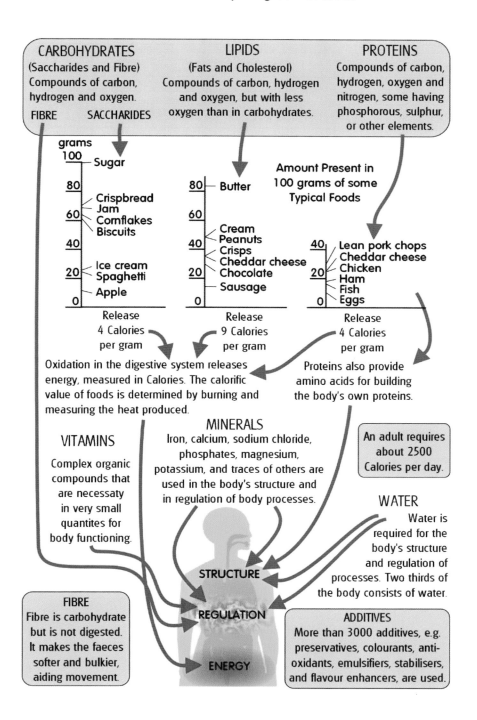

54. Food

55. Human Metabolism

A degree of digestion of food takes place in the mouth and stomach but most occurs in the small intestine. The products of digestion are passed to the blood stream, some via the liver where further processing takes place.

The bloodstream carries the food products, and oxygen from the lungs, to the cells in all parts of the body.

In the cell, glucose is converted by a series of steps to pyruvate, which is passed to the mitochondrion. The mitochondrion is a specialised unit in the cell that produces most of the cell's energy requirements in the form of adenosine triphosphate (ATP). Some ATP is produced in the earlier glucose conversion.

ATP is the energy source essential for the metabolism of animals and plants. It provides energy by reverting to phosphate and adenosine diphosphate (ADP).

The mitochondrion takes in the pyruvate and oxygen, and passes water and carbon dioxide back to the bloodstream.

The cell uses the energy to build proteins, transport materials and power muscular action.

Excess intake of food is dealt with by storage as glycogen and fat. At the other extreme, if glucose is in short supply for conversion to pyruvate, glycogen is used, then fat and lastly protein from body muscle.

56. Reproduction   58. The Endocrine System
         59. The Nervous System

How Everything Came About

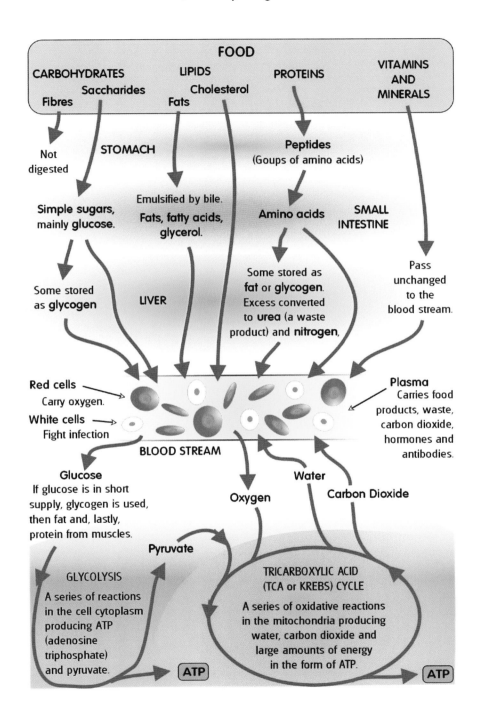

55. Human Metabolism

56. Reproduction

Mitosis is the division of a cell into two and is the means of replacing dead cells and providing growth and development of the body. Meiosis is the coming together of cells from male and female to produce offspring.

Each cell in the body contains 46 chromosomes, 23 from each parent. Each chromosome consists of two identical chromatids and each chromatid consist of a molecule of DNA. The DNA is like a twisted ladder, each 'rung' consisting of two molecules paired in a specific way. A gene is a section of the DNA consisting of a particular sequence of pairs. The full set of 32,000 genes controls characteristics of the person.

When a cell divides into two, in mitosis, copies of the chromatids are enclosed in each new cell.

In meiosis, the genes from each parent are randomly exchanged in the chromosomes of the cells required for sexual reproduction. Two cell divisions then produce four cells, each with one chromatid that later duplicates to become a chromosome. These cells, spermatozoa in the male and ova in the female, contain only23 chromosomes.

On mating the male and female cells form a zygote containing 46 chromosomes, 23 from each parent. The zygote grows by mitosis to produce an embryo.

57. Genetics

How Everything Came About

Cells reproduce by mitosis for growth and to replace worn out cells. Egg and sperm cells for sexual reproduction are produced by meiosis. Both processes involve chromosomes.

CHROMOSOMES

In the nucleus of each cell of the human body are 46 chromosomes, 23 from each parent. Two joined identical chromatids make up each chromosome. One pair of chromosomes is shown.

From father
From mother

DNA

A chromatid is a molecule of deoxyribonucleic acid coated with protein. It is like a twisted ladder, each 'rung' consisting of a pair of molecules. The DNA in the 23 chromosomes has 3 billion base pairs and is nearly one metre long. A gene is a section of a chromatid having a particular sequence of base pairs to code the activity of the cell.

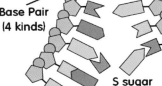

In reproduction, the DNA splits and each half adds molecules to form a replica.

Base Pair (4 kinds)

S sugar
P phosphate
A adenine
T thymine
C cytosine
G guanine

MITOSIS

The two chromatids separate and replicas are produced.

The cell divides producing two identical cells.

The 23rd pair of chromosomes are matched only in females (XX). Males have one female and one male chromosome (XY). Thus the spermatozoa can be male or female.

MEIOSIS

The matching pairs of chromatids come together and randomly exchange sections, ie, genes.

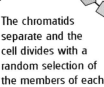

The chromatids separate and the cell divides with a random selection of the members of each pair.

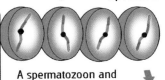

The two cells divide producing 4 gametes (spermatozoa in the male, ova in the female), each with 23 chromosomes (single chromatids at this stage) consisting of a random mix of genes from both parents.

A spermatozoon and an ovum combine in sexual reproduction to form a zygote, which multiplies by mitosis to produces an embryo.

56. Reproduction

57. Genetics

Every living cell in the body carries a copy of the individual's genes, but the cell uses only the genes required for its particular tasks. The cell has, in fact, two corresponding sets of genes, one from each parent, and the dominant gene is used. The recessive one is ignored, though it will be passed to offspring.

An essential role of the cell is to build proteins from amino acid molecules. The assembly unit, the ribosome, moves along a replica of the section of DNA representing the appropriate gene. The sequence of molecules in the gene identifies the required amino acid. The amino acid molecule is moved into position. This repeats until the protein molecule is complete.

All of the 32,000 human genes have been identified and mapped in the chromosomes. Determining what each gene does is progressing. Each individual's DNA sequence is unique and can be used for identification.

Genetic engineering is used for drug production, replacement of defective genes, and improvement of crops. Cloning of animals has been achieved but the possibility of cloning humans is controversial.

Damage to the genes, or copying errors, can be disruptive or fatal to the cell, but can introduce beneficial evolutionary trends.

A gene is a segment of chromosome which controls a characteristic of the body by carrying instructions for the making of a particular protein. Every cell contains every gene but only the genes required for the particular type of cell become active.

MECHANISM OF GENE ACTION

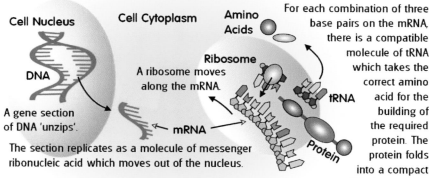

Cell Nucleus — DNA. A gene section of DNA 'unzips'. The section replicates as a molecule of messenger ribonucleic acid which moves out of the nucleus.

Cell Cytoplasm — A ribosome moves along the mRNA.

For each combination of three base pairs on the mRNA, there is a compatible molecule of tRNA which takes the correct amino acid for the building of the required protein. The protein folds into a compact 3D shape which determines its correct functioning. Many diseases are due to incorrect folding.

The cell has a set of chromosomes from each parent, so pairs of genes control each characteristic. A dominant gene controls, if paired with a recessive gene, e.g. dark hair is dominant over red hair. Recessive genes may still be passed on and affect future generations.

GENETIC ENGINEERING

Genes can be transferred from one animal or plant into another. Insulin, once only available from the pancreas of animals, is now obtained by injecting the human insulin gene into bacteria. The modified bacteria produce unlimited quantities.

CLONING

Production of genetically identical plants is easy, by taking cuttings, for example. It is more difficult with animals but many different kinds have been cloned by exchanging cell nuclei. Some instances of cloning of endangered and extict species have been achieved. The possibility of cloning humans raises ethical issues and is the subject of much debate.

MUTATIONS

Damaging radiation and some chemicals can cause changes in the genetic code. The changes are usually fatal to the cell or not beneficial. If beneficial, they can improve survival and contribute to evolution.

GENETIC FINGERPRINTS

Although 99.9% of the DNA in humans is the same, this still leaves 3 million base pairs that can differ between individuals. Specific markers in the DNA sequence can thus be used to identify individuals.

HUMAN GENOME PROJECT

The 3 billion base pairs in the 23 chromosomes have been mapped and the 32,000 human genes identified. The number of genes required for each organ has been roughly determined and the genetic code for several hereditary diseases is known.

55. Human Metabolism

58. The Endocrine System

The endocrine system is the body's chemical control system. It is slower acting than the nervous system which is the body's electrical control system.

Glands located in various parts of the body produce hormones. These are chemicals that perform specific tasks, such as the regulation of blood composition, sexual characteristics or the functioning of sex organs.

The control of the endocrine glands is not entirely centralised but nearly so. The hypothalamus is a region of the brain that is adjacent to and controls the pituitary gland. This gland supplies a number of hormones that act directly, but also provides hormones that control other glands.

The hormones are released into the blood stream for circulation, but are used only by cells that require them.

How Everything Came About

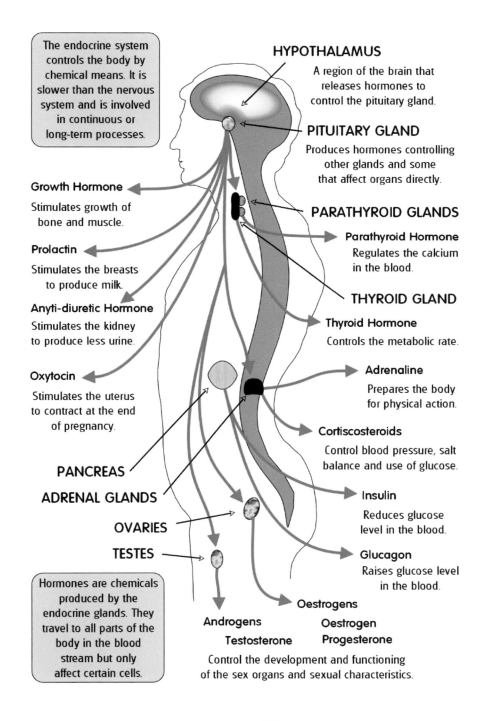

How Everything Came About

55. Human Metabolism

59. The Nervous System

The nervous system, extending from the brain to all parts of the body, consists of neurons. A neuron is a cell which has a fibrous extension, a nerve fibre. Some of the nerve fibres are more than a metre long.

Sensory neurons connect to sense organs while motor neurons connect to muscles and glands.

The brain and spinal cord constitute the central nervous system, and house intermediate neurons which connect neurons to other neurons. In this way, sensory events are recognised and muscles or glands prompted to respond.

When the sense organ produces a stimulus, a nerve impulse, which is electrical, travels along the nerve fibre. At the junctions between neurons there is a small gap, a synapse. The impulse travels across the gap, not electrically but chemically by transmitter substances diffusing across the gap.

60. The Senses

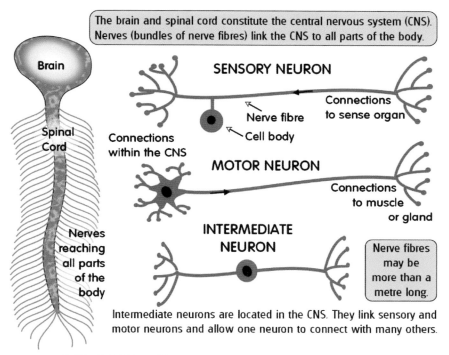

The brain and spinal cord constitute the central nervous system (CNS). Nerves (bundles of nerve fibres) link the CNS to all parts of the body.

Intermediate neurons are located in the CNS. They link sensory and motor neurons and allow one neuron to connect with many others.

NERVE IMPULSES

A stimulus creates an electrical pulse which travels along the fibre.

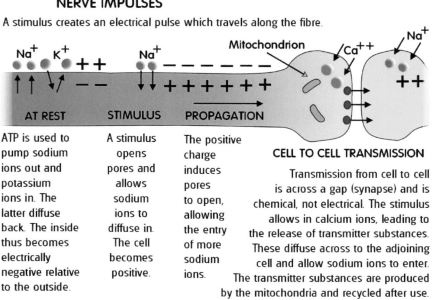

AT REST
ATP is used to pump sodium ions out and potassium ions in. The latter diffuse back. The inside thus becomes electrically negative relative to the outside.

STIMULUS
A stimulus opens pores and allows sodium ions to diffuse in. The cell becomes positive.

PROPAGATION
The positive charge induces pores to open, allowing the entry of more sodium ions.

CELL TO CELL TRANSMISSION
Transmission from cell to cell is across a gap (synapse) and is chemical, not electrical. The stimulus allows in calcium ions, leading to the release of transmitter substances. These diffuse across to the adjoining cell and allow sodium ions to enter. The transmitter substances are produced by the mitochondria and recycled after use.

59. The Nervous System

60. The Senses

The sense organs contain receptor cells which respond to external stimuli by triggering impulses in sensory neurons. The brain interprets the signals as vision, sound, touch, etc., according to the area of the brain where they are received.

There are several types of receptor cells.

Photoreceptors respond to light and are found in the retina of the eye.

Mechanoreceptors respond to movement and are located in the ears to detect sound, balance and the direction of gravity. They are also located in the skin to detect touch and pressure.

Chemoreceptors detect and distinguish particular substances by taking them into solution in the mouth and nose for interpretation as taste and scent.

Thermoreceptors in the skin detect warmth and cold.

Nociceptors are complex receptors that can detect noxious substances. They respond to mechanical, chemical or thermal stimuli.

In addition to external sensors, there are internal ones that monitor the state of muscles and limbs and regulate body processes.

61. The Brain

How Everything Came About

Receptor cells, when stimulated, transmit impulses to the nervous system. Depending on the region of the brain receiving the impulses, they are interpreted as vision, etc.

EYE — Optic Nerve Connects to the brain.

Lens Focusses an image on the retina.

Retina Contains rods and cones.

PHOTORECEPTORS

SIGHT
Light photons split the pigment molecules, generating an electrical impulse in the neuron. ATP reforms the molecules.

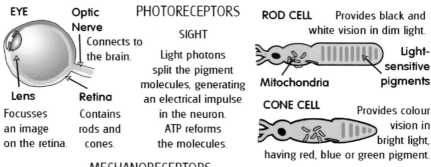

ROD CELL Provides black and white vision in dim light.

Light-sensitive pigments

Mitochondria

CONE CELL Provides colour vision in bright light, having red, blue or green pigment.

MECHANORECEPTORS

The membranes of the receptor cells stretch. Their conductance changes and allows ions to flow, sending impulses to the neurons.

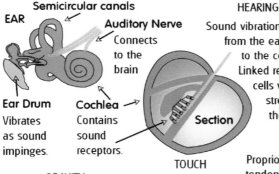

Semicircular canals

EAR

Auditory Nerve Connects to the brain

Ear Drum Vibrates as sound impinges.

Cochlea Contains sound receptors.

Section

HEARING
Sound vibrations pass from the ear drum to the cochlea. Linked receptor cells vibrate, stretching the links.

BALANCE
Three semicircular canals, at right angles to each other, contain fluid which resists sudden motion of the head. Receptor cells, one end embedded in the fluid, are stretched.

GRAVITY
Receptor cells in the ear stretch as the head tilts and sense the direction of gravity.

TOUCH
Receptors in the skin sense touch and pressure.

INTERNAL SENSING
Proprioceptors, in muscles, ligaments, tendons and joints, transmit details of the state of the limbs and muscles.

CHEMORECEPTORS

Protein molecules in the outer membranes of the receptor cells fit, in terms of chemistry and shape, with specific substances, and change the membrane conductance.

THERMORECEPTORS
Separate receptors in the skin sense warmth and cold.

NOCICEPTORS
Some sensors of noxious stimuli, and itching sensors, respond to mechanical, chemical or thermal signals.

TASTE AND SCENT
Substances dissolve in saliva and mucus.

INTERNAL SENSING
A network of receptors allows regulation of body processes.

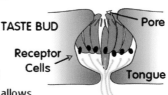

Pore

TASTE BUD

Receptor Cells

Tongue

Nerve Fibre

60. The Senses

61. The Brain

The brain consists of about 100 billion neurons each having up to 10,000 connections to other neurons. It is thus an enormous network with parallel connections, but has no identifiable central control unit.

Particular areas of the brain are involved with particular sensations and activities, and the left half of the brain controls the right side of the body and vice versa.

Because the network consists of parallel connections, different processing tasks can be undertaken simultaneously. This is not so in a computer where steps are sequential.

The connections between the neurons are not permanent. New connections are made and others severed. This ability to vary the circuitry is probably related to the properties of learning and memory. Parallel paths can be strengthened if useful and abandoned if not, giving a learning progression. Memory, rather than being localised, is more likely to be embodied in circulating signals, the stronger the paths the better the recollection.

The source and mechanism of consciousness and self-awareness present outstanding problems. Science can offer little by way of explanation. Some believe that quantum mechanics may provide answers.

How Everything Came About

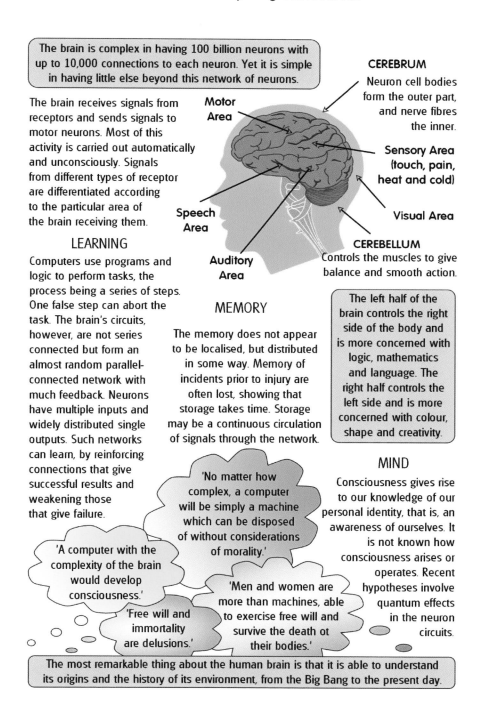

The brain is complex in having 100 billion neurons with up to 10,000 connections to each neuron. Yet it is simple in having little else beyond this network of neurons.

The brain receives signals from receptors and sends signals to motor neurons. Most of this activity is carried out automatically and unconsciously. Signals from different types of receptor are differentiated according to the particular area of the brain receiving them.

CEREBRUM
Neuron cell bodies form the outer part, and nerve fibres the inner.

Motor Area

Sensory Area (touch, pain, heat and cold)

Speech Area

Visual Area

Auditory Area

CEREBELLUM
Controls the muscles to give balance and smooth action.

LEARNING

Computers use programs and logic to perform tasks, the process being a series of steps. One false step can abort the task. The brain's circuits, however, are not series connected but form an almost random parallel-connected network with much feedback. Neurons have multiple inputs and widely distributed single outputs. Such networks can learn, by reinforcing connections that give successful results and weakening those that give failure.

MEMORY

The memory does not appear to be localised, but distributed in some way. Memory of incidents prior to injury are often lost, showing that storage takes time. Storage may be a continuous circulation of signals through the network.

The left half of the brain controls the right side of the body and is more concerned with logic, mathematics and language. The right half controls the left side and is more concerned with colour, shape and creativity.

'A computer with the complexity of the brain would develop consciousness.'

'No matter how complex, a computer will be simply a machine which can be disposed of without considerations of morality.'

'Free will and immortality are delusions.'

'Men and women are more than machines, able to exercise free will and survive the death of their bodies.'

MIND

Consciousness gives rise to our knowledge of our personal identity, that is, an awareness of ourselves. It is not known how consciousness arises or operates. Recent hypotheses involve quantum effects in the neuron circuits.

The most remarkable thing about the human brain is that it is able to understand its origins and the history of its environment, from the Big Bang to the present day.

How Everything Came About

1. The First 400.000 Years

62. Quantum Mechanics

Photons behave as both particles and waves, though it is not possible to visualise how this could be. Just as surprising is the fact that small particles with mass can behave as waves. Large objects would also behave as waves but their wavelengths are too small to allow any observable effects.

The proper description of a particle is a wave packet which has a spatial distribution. Certain properties of the particle, such as position and velocity are not definite, but can be expressed only as probabilities. The more definite is its position, the more uncertain is its velocity and vice versa. Moreover, some properties are not definite until measured but exist in a superposition of possible states.

The interaction between particles is properly described by combining the mathematical descriptions of the wave packets. In this way the whole of chemistry could, in principle, be explained, though in practice the complexity is great.

Energy is available only in tiny discrete quanta and cannot be subdivided indefinitely. This gives rise to energy levels in atoms to which the electrons, protons and neutrons are restricted. Without these quantised energy levels atoms would not be stable.

63. String Theory

WAVE NATURE OF PARTICLES

Not only do photons have the properties of both particles and waves, but all particles exhibit wave properties. Electron beams, e.g. are used in electron microscopes. Even a golf ball has wave properties but because its mass is large its wavelength is tiny and can never be observed.

Photon

Electron (wavelength shorter than visible light photon)

Golf ball (wavelength negligible)

HEISENBERG'S UNCERTAINTY PRINCIPLE

The more certain the position of a particle becomes, the more uncertain is its velocity, and vice versa. Similarly, the more certain its energy, the less certain the time period during which it has the energy. The uncertainties are not due to inadequecies in the ability to measure but due to uncertainties in the way things are.

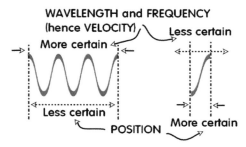

SCHRODINGER'S WAVE EQUATION

The proper description of a particle is by a mathematical equation that explains the particle as a wave packet, and determines particle properties in terms of probabilities of particular values being observed.

UNIFICATION

Quantum mechanics successfully describes the structure and behaviour of all material and radiation in the universe, but does not incorporate the theory of gravity. String theory offers one possibility of uniting quantum mechanics and gravity.

CHEMISTRY

Molecules form by the interaction of outer electrons of atoms. Quantum mechanics can thus explain chemical processes, but the calculations are difficult and have been completed for simple situations only.

ENERGY QUANTA AND THE ELEMENTS

Energy is available only as tiny discrete quanta. This restricts the values of energy of particles in atoms, and results in energy levels that protons, neutrons and electrons must occupy. This accounts for the structure of all the elements and isotopes.

'BORROWED' ENERGY

Particles come into existence in empty space for a very short time, the shorter the time the greater their energy. Similarly, particles can 'borrow' energy for a short time.

62. Quantum Mechanics

63. String Theory

String theory extends quantum mechanics and is fully consistent with it. Its advantage is that it can incorporate gravity whereas, currently, it is not possible to bring quantum mechanics and gravity into a unified theory. There is however no experimental confirmation of string theory.

Elementary particles, according to string theory, consist of tiny vibrating lengths of one-dimensional string. Particle properties such as mass and electrical charge arise from the vibrational patterns. One of the patterns has the expected properties of the graviton, the carrier particle of gravitational waves.

There have been many versions of string theory. One of them, M-theory, was recognised as being consistent with previous theories.

All string theories require there to be more than three dimensions of space. M-theory requires 10 space dimensions. We are aware of only three, so where are the others? It is suggested that they are curled up very small at every point in space. Had they not been curled up, the Universe could not have developed the way it did. With four macroscopic dimensions, for example, the orbits of planets would not be stable.

How Everything Came About

> String theory proposes that elementary particles consist of tiny vibrating lengths of one-dimensional string. The vibrations are analogous to those on a violin string but the strings can also be endless loops. The different vibrational patterns give rise to the particle properties such as mass and charge.

String length is about 1 (20 zeros) times smaller than an atom nucleus.

The theory is fully consistent with quantum theory but has the advantage of incorporating gravity, one vibrational pattern having exactly the expected properties of the graviton.

The theory requires that the universe has more than three dimensions, to provide the required pattern of vibrational modes. A violin string can vibrate in only two dimensions though there are an infinite number of overtones in addition to the fundamental frequency.

The additional dimensions that we are not aware of are considered to be curled up and very small, so the distance that could be travelled before coming back to the start is smaller than could ever be measured. As we sweep a hand through the air, each part of the hand traverses each dimension at each point in space, and returns to where it was.

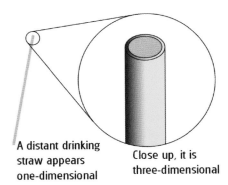

A distant drinking straw appears one-dimensional. Close up, it is three-dimensional.

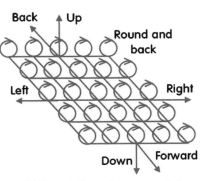

Additional dimensions are curled up at every point in space.

WHY ARE JUST THREE DIMENSIONS APPARENT?

Complex structures could hardly exist with less than three extended spatial dimensions. Problems arise with more than three. For example, with four dimensions or more, orbits of planets, etc., would not be stable.

M-THEORY

M-theory is consistent with previous string theories and contains, not only strings, but vibrating two-dimensional membranes, three-dimensional blobs and many other features. It requires 11 dimensions (10 space and 1 time).

62. Quantum Mechanics

64. Quantum Entanglement

When particles transform to other particles, properties such as energy, spin, and electrical charge have in total to remain the same. For example, if a pair of electrons whose total spin must be zero are produced, the spin of one will be in the opposite direction to the spin of the other. Spin directions are called up and down, rather than clockwise or anticlockwise.

However, neither spin will be in any particular direction until it is measured. A measurement on one electron instantaneously fixes the spin of its partner, no matter how far apart they are. The spin of the two is said to be entangled. How the link is maintained between the entangled particles is not understood: it is one the amazing aspects of quantum mechanics.

Entanglement has practical applications in quantum cryptography and, in principle, teleportation. The fact that spin direction is indeterminate until measured is allowing the development of quantum computing.

A further consequence of quantum mechanics arises in the measurement of the spin of a particle. Because spin is quantised and indeterminate in direction until measured, the measurement will always show it to be in line with the measurement direction or exactly opposite (up or down). The spin will never be found to be at an angle to the measuring direction.

65. Quantum Computing

How Everything Came About

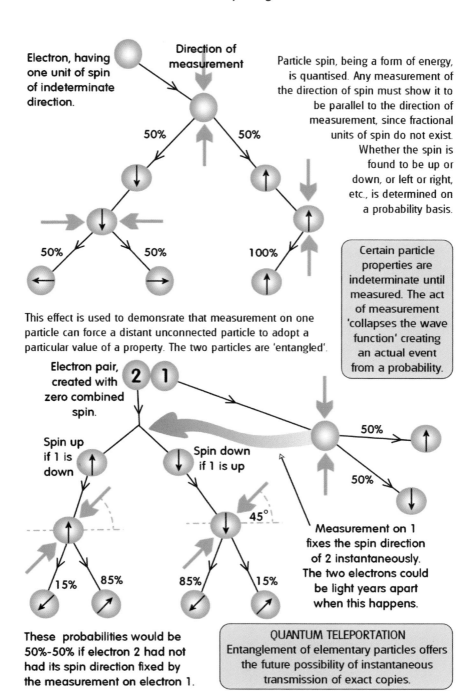

Electron, having one unit of spin of indeterminate direction.

Direction of measurement

Particle spin, being a form of energy, is quantised. Any measurement of the direction of spin must show it to be parallel to the direction of measurement, since fractional units of spin do not exist. Whether the spin is found to be up or down, or left or right, etc., is determined on a probability basis.

This effect is used to demonsrate that measurement on one particle can force a distant unconnected particle to adopt a particular value of a property. The two particles are 'entangled'.

Certain particle properties are indeterminate until measured. The act of measurement 'collapses the wave function' creating an actual event from a probability.

Electron pair, created with zero combined spin.

Spin up if 1 is down

Spin down if 1 is up

Measurement on 1 fixes the spin direction of 2 instantaneously. The two electrons could be light years apart when this happens.

These probabilities would be 50%-50% if electron 2 had not had its spin direction fixed by the measurement on electron 1.

QUANTUM TELEPORTATION
Entanglement of elementary particles offers the future possibility of instantaneous transmission of exact copies.

64. Quantum Entanglement

65. Quantum Computing

Computers process and store information using binary mathematics. In binary there are only two digits, 0 and 1, so they can easily be represented by a yes/no device. A switch can be open or closed, a space on a CD track can be filled in or left blank. One unit of information, 0 or 1, is called a bit.

Quantum computing is based on the fact that particles can exist in a superposition of two different states at the same time. This is analogous to a switch that can be in an open and a closed state at the same time. With two such switches there are four possible combinations of open and closed, so four bits of information can be processed simultaneously or stored, rather than two. As more such switches are added the number of combinations increases rapidly.

The superposition of 0 and 1 is called a qubit. More than a million different numbers can be stored by as few as 20 qubits.

The technology is at an early stage and only a few trivial calculations have been achieved so far. One major problem is isolation from the environment of the particles in the superposed state. Interaction with the environment causes decoherence, that is, a collapse to one or other of the two states.

66. Quantum Cryptography

How Everything Came About

One bit (yes/no) of information can be stored by, for example, a switch (on/off), a capacitor (charged/discharged) or electron spin (up/down). One of two binary numbers can thus be stored. In a conventional computer each bit has to be stored separately and the bits processed one at a time.

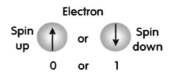

Electron — Spin up or Spin down — 0 or 1

Before the spin of an electron, or other elementary particle, is measured it is indeterminate. There is a superposition of up and down states, so two numbers can be stored. This is called a qubit.

0
1

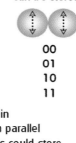

With 2 electrons, 4 binary numbers can be stored.

00
01
10
11

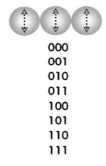

With 3 electrons, 8 binary numbers can be stored, and so on.

000
001
010
011
100
101
110
111

A quantum computer, using qubits, would exhibit remarkable speed increase over a conventional computer in being able to process many numbers in parallel simultaneously. For example, 100 qubits could store 1,268,000,000,000,000,000,000,000,000,000 numbers that could be processed simultaneously.

By using entanglement, logic gates can be set up for mathematical operations

> Quantum technology will provide a means of code cracking, but it has already provided uncrackable codes (quantum cryptography).

APPLICATIONS

A feasible major application is in determining the factors of large numbers. For example,
127 x 129 = ? is easy to solve but
29083 = ? x ? takes a lot longer.
Commonly used codes employ factors of large numbers because computers cannot factorise quickly. A thousand digit number would take longer than the age of the universe. A quantum computer would crack these codes readily. Also, searching a database of N items could be done in square root of N operations, instead of half N.

PROBLEMS

The qubits will readily interact with the environment and collapse to one or other of their two states (decoherence). As the number of qubits is increased, the problem gets more difficult to control. The use of quantum computers may always be restricted to solving specific mathematical problems.

PROGRESS

Quantum computers are at an early stage of devolpment as evidenced by the variety of systems being explored and developed. These include the spin of electrons, nuclei, atoms and molecules. Experimental computers have been produced and used to solve fairly trivial arithmetic problems, but versatile commercially available quantum computers are for the future.

65. Quantum Computing

66. Quantum Cryptography

The polarisation of a photon is analogous to the spin of an electron, or other particle having mass. Polarisation is thus quantised and will always be found to be in line with the direction of measurement or at right angles to it. A photon already polarised at 45 degrees to the measuring direction will, with equal probability, be found to be inline or at right angles to the measuring direction.

This has allowed the development of a secure means of transmitting coded messages. A stream of photons, each polarised in a random but recorded direction, is sent via an optical fibre. Distances of over 100 km have been achieved.

At the receiver, the polarisation direction of each is measured, using random directions of measurement. The directions of polarisation and measurement are then compared {publicly) to eliminate incompatible combinations. The results of measurements from the remaining compatible combinations provide a secure set of data fora code key.

Any attempted interception destroys the photon, which cannot be copied because its original polarisation direction cannot be determined.

Advanced methods with increased security involve sending one of a pair of entangled photons.

How Everything Came About

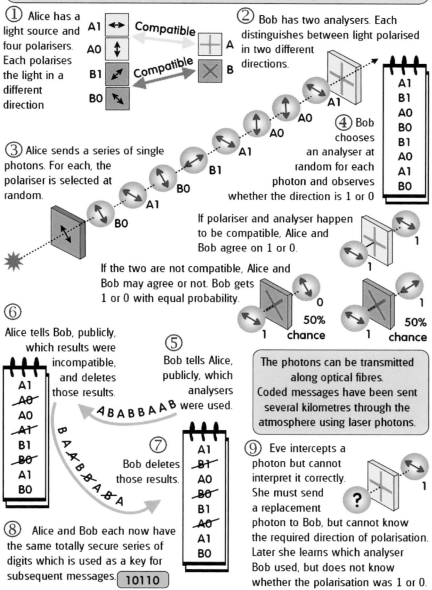

1. The First 400,000 Years

67. General Relativity

General relativity is essentially the theory of gravity. The force of gravity causes bodies to accelerate just as mechanical, electrical, etc. forces do. There is no difference between experiencing a gravitational field and an acceleration. It follows that light bends in a gravitational field. Moreover, a gravitational field due to a mass is a bending of space by the presence of the mass.

More correctly it is a warping of four-dimensional spacetime, and the rate of flow of time is affected by the field. Time runs slower and lengths become shorter the stronger the gravitational field. These changes are relative to observers in weaker fields and are not apparent to those experiencing the changes. Time runs faster for global positioning satellites (GPS) at 45 microseconds per day. A correction has to be made as the error would accumulate.

A black hole is an accumulation of so great a mass that nothing not even light, can escape. Space is warped back on itself.

Gravitational waves travelling through space are carried by massless particles, gravitons, which have yet to be detected. The first gravitational wave to be detected was the result of a collision of two black holes.

68. Special Relativity

How Everything Came About

Einstein's General Relativity theory concerns objects moving with acceleration. Because of the Equivalence Principle, the theory provides the theory of gravity.

EQUIVALENCE PRINCIPLE

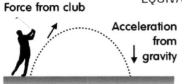

Force from club
Acceleration from gravity

Objects are accelerated by applied forces. Gravity produces acceleration and it is impossible to tell the difference between acceleration from an applied force and the presence of a gravitational field. The two are equivalent.

In a motionless space ship, away from gravitational fields, objects are weightless.

If the ship accelerates the weight extends the spring and a beam of light entering the ship appears to bend as it passes across.

When the ship is stationary in a gravitational field the effects are identical. Thus gravity bends a beam of light, albeit very slightly.

Earth

ESCAPE VELOCITY

For a rocket to escape from Earth, its velocity must be sufficient to allow for the deceleration due to gravity. The gravity of a black hole is so strong that the escape velocity is greater than the speed of light. Photons cannot escape.

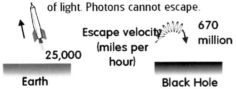

25,000   Escape velocity (miles per hour)   670 million

Earth   Black Hole

SPACE WARPING

Objects with mass warp space (more correctly, four-dimensional spacetime). Light takes the shortest path through warped space. A black hole warps space back on itself.

TIME DILATION

Clocks on an accelerating space ship, front and back, record different times. Synchronising pulses from the front clock arrive at the back clock closer together (Doppler effect) because the back is moving faster when they arrive. So the back clock is seen to run slower than the front one. Similarly, clocks run slower in a gravitational field. As time slows down, lengths shorten. In a black hole time stops, but only as observerd from outside.

GRAVITATIONAL WAVES

Moving masses send ripples through space at the speed of light. The waves are weak and difficult to detect. The carrier particle is the graviton, which has no mass. It acts as the photon does for electromagnetic waves.

Rotating Binary Star System   Gravitational Wave

67. General Relativity

68. Special Relativity

Velocity is relative, meaning that we can experience velocity only relative to something that is changing its position from us as time progresses. Light, however, behaves strangely. No matter how fast a source of light is moving relative to us, we still observe the light to be moving at a constant velocity.

The consequences of this are contrary to expectation. If we observe travellers moving relative to us, we see that their time passes more slowly than our time. Their measures of length, relative to ours, are shortened in the direction of motion. Equally, however, they observe our time to be running slower than theirs and our measures of length to be shortened. We observe their mass to increase, they observe our mass to increase.

In addition, simultaneity has no precise meaning. Events that are simultaneous for one observer may be sequential for another.

A further consequence is that mass and energy are equivalent and related by Einstein's famous $E=mc^2$ equation, where c is the velocity of light.

Because of their velocity, global positioning satellites (GPS) experience a slowing of time of 7 microseconds per day. This would accumulate giving serious errors and has to be corrected for.

69. The Twin Paradox

How Everything Came About

> Einstein's Special Relativity theory applies to objects moving relative to each other at constant velocity. Two observers in relative motion each sees the other's clocks to run slow, lengths to be shortened in the direction of motion, and masses to increase. The effect is negligible unless the velocity is an appreciable fraction of the velocity of light.

A space ship leaves earth and travels at a speed of 60% of the velocity of light, i.e. at 0.6 light years per year. It passes a space colony located 12 light years from Earth.

EARTH'S LOG
Distance 12 light years.
Speed 0.6.
Time 12/0.6 = 20 years.
Ship's clock seen to be running 20% slow, i.e. 16 years for 20 Earth years.
Ship's distances seen to be 20% shorter, i.e. 9.6 light years for 12 Earth light years.

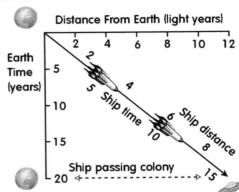

From the viewpoint of the moving ship, Earth is receding, and the colony approaching, at a speed of 0.6. But the distance to the colony and the journey time are now found to be shorter.

The changes in time, length and mass arise because the velocity of light is the same for all observers. They are calculated by the Lorentz equations.

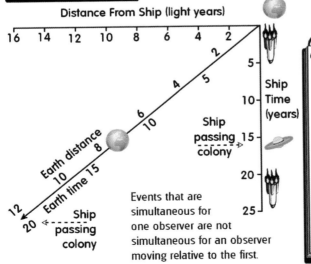

SHIP'S LOG
Time 16 years.
Speed 0.6.
Distance 16 x 0.6 = 9.6 light years.
Earth's clock seen to be running 20% slow, i.e. 12.8 years for 16 Ship years. Earth's distances seen to be 20% shorter, i.e. 7.68 light years for 9.6 Ship light years.

Events that are simultaneous for one observer are not simultaneous for an observer moving relative to the first.

> The equivalence of mass and energy ($E=mc^2$) is a consequence of Special Relativity.

68. Special Relativity

69. The Twin Paradox

The so-called 'twin paradox' is not a paradox but a well-established consequence of special relativity. One of a pair of twins stays on Earth while the other travels in space at a velocity which is an appreciable fraction of the speed of light. The travelling twin returns younger than the twin who remained on Earth, because of the time dilation effect.

The hypothetical situation was referred to as a paradox and created some debate, because, velocity being relative, it was claimed that the final situation should be the same for each twin. However, the travelling twin has to decelerate and accelerate in order to join a returning ship. This ship already has time shifts relative to Earth, because of its velocity relative to Earth. Thus the situation is not symmetrical, and the travelling twin is the one who ends up younger on Earth.

Journeys at such velocities are of course beyond what is currently possible. Nevertheless, the time shift has been demonstrated many times by measuring the lifetimes of particles travelling at high velocities, and by observing extremely accurate clocks for long periods on aircraft.

How Everything Came About

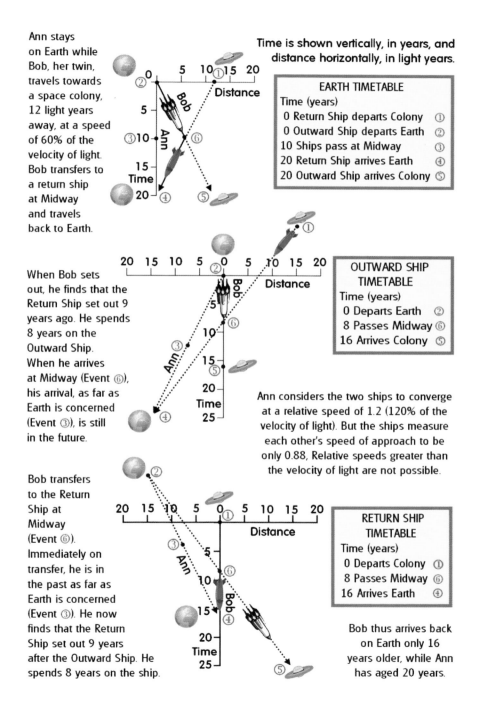

Ann stays on Earth while Bob, her twin, travels towards a space colony, 12 light years away, at a speed of 60% of the velocity of light. Bob transfers to a return ship at Midway and travels back to Earth.

Time is shown vertically, in years, and distance horizontally, in light years.

EARTH TIMETABLE
Time (years)
0 Return Ship departs Colony ①
0 Outward Ship departs Earth ②
10 Ships pass at Midway ③
20 Return Ship arrives Earth ④
20 Outward Ship arrives Colony ⑤

When Bob sets out, he finds that the Return Ship set out 9 years ago. He spends 8 years on the Outward Ship. When he arrives at Midway (Event ⑥), his arrival, as far as Earth is concerned (Event ③), is still in the future.

OUTWARD SHIP TIMETABLE
Time (years)
0 Departs Earth ②
8 Passes Midway ⑥
16 Arrives Colony ⑤

Ann considers the two ships to converge at a relative speed of 1.2 (120% of the velocity of light). But the ships measure each other's speed of approach to be only 0.88. Relative speeds greater than the velocity of light are not possible.

Bob transfers to the Return Ship at Midway (Event ⑥). Immediately on transfer, he is in the past as far as Earth is concerned (Event ③). He now finds that the Return Ship set out 9 years after the Outward Ship. He spends 8 years on the ship.

RETURN SHIP TIMETABLE
Time (years)
0 Departs Colony ①
8 Passes Midway ⑥
16 Arrives Earth ④

Bob thus arrives back on Earth only 16 years older, while Ann has aged 20 years.

1. The First 400,000 Years

70. Why is the Universe as it is?

The Universe has evolved to its present state because of a large number of circumstances. Had many of these been very slightly different there would have been no possibility of a Universe supporting life, or even anything resembling our Universe.

Had there not been a slight excess of matter over antimatter (one part in billion) shortly after the Big Bang there would have simply been expanding radiation. If the gravitational force, the expansion rate and the electrical force had not been finely balanced the Universe would have either consisted entirely of expanding gas and radiation or would have rapidly collapsed back to its tiny origin.

Put simply, there have been many coincidences that have led to our Universe in is present state. We can say, of course, that if the Universe had not developed in the way it has, we would not be here to comment on it and express surprise that we are here. This is the basis of the Anthropic Principle which denies that we should be surprised by the string of coincidences.

Those who speculate about parallel universes or a series of previous universes can of course argue that our Universe may simply be one of many that just by chance happened to have the particular combination of circumstances that we observe.

71. Is the Future Determined?

How Everything Came About

Design?

The present state of the universe is very finely balanced. Had conditions been slightly different immediately after the Big Bang, a complex universe capable of giving rise to intelligent life forms would never have developed.

Chance?

A few examples of the consequences of very slight changes are shown.

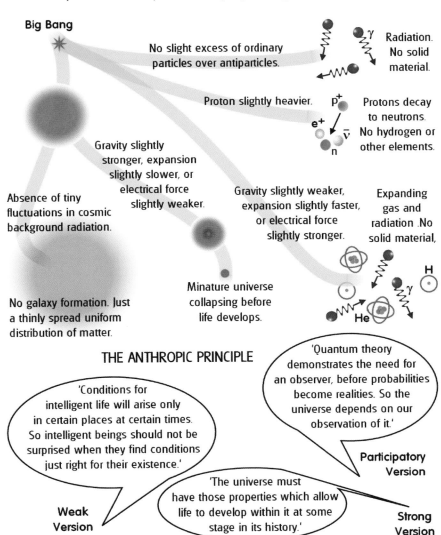

Big Bang

No slight excess of ordinary particles over antiparticles. → Radiation. No solid material.

Proton slightly heavier. → Protons decay to neutrons. No hydrogen or other elements.

Gravity slightly stronger, expansion slightly slower, or electrical force slightly weaker.

Gravity slightly weaker, expansion slightly faster, or electrical force slightly stronger. → Expanding gas and radiation. No solid material.

Absence of tiny fluctuations in cosmic background radiation.

Minature universe collapsing before life develops.

No galaxy formation. Just a thinly spread uniform distribution of matter.

THE ANTHROPIC PRINCIPLE

'Conditions for intelligent life will arise only in certain places at certain times. So intelligent beings should not be surprised when they find conditions just right for their existence.'
Weak Version

'The universe must have those properties which allow life to develop within it at some stage in its history.'
Strong Version

'Quantum theory demonstrates the need for an observer, before probabilities become realities. So the universe depends on our observation of it.'
Participatory Version

70. Why is the Universe as it is?

71. Is the Future Determined?

In the early years of the 20th century the Universe was visualised as a large clockwork-like mechanism. Since every particle had a particular motion, its future interactions with all other particles were determined and thus the entire future of everything and everybody was determined.

The discovery of quantum mechanics put an end to this deterministic view. It was found that many particle properties exist as probabilities rather than definite values.

Randomness exists in nature in an intrinsic way and is not simply a manifestation of our inability to sort out what is really happening. More often than not, of course, random events average out so that useful, and very accurate, predictions can be made. But this is not always so.

Some sequences of events, called chaotic, are determined exactly but can never be predicted exactly, because the outcomes are too sensitive to tiny variations in the initial conditions. Weather forecasting is in this category.

How Everything Came About

DETERMINISM

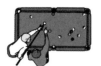

Once the cue ball moves, the future positions of all the balls are determined.

Since every particle in the universe has a precise position and motion, all collisions and interactions in the future are precisely determined. Thus all events are predestined. This argument is no longer valid.

The universe is not a giant clockwork mechanism.

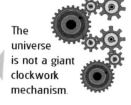

CHAOS THEORY

Many systems settle down and remain unaffected by a slight disturbance. But some develop along an entirely different course. Weather is an example. These chaotic systems are deterministic but their future states could never be predicted.

A butterfly flapping its wings in Brazil may cause a tornado in Texas.

QUANTUM MECHANICS

The properties of particles exist only as probabilities. Actual properties appear only when observations are made. As a property is observed more precisely, greater uncertainty results in other properties.

We picture electrons orbiting the nucleus of an atom. In fact, there is no more than a probability of finding an electron in a particular location, the probability being greater for some locations than others.

RANDOM EVENTS

Commonly experienced events are often predictable because they depend on average properties of the constituent particles. But in some situations random events arising from the intrinsic probabilities in nature can be easily seen.

The temperature of a liquid when heated can be predicted because it is the average energy of the liquid molecules that determines the temperature.

Two identical radioactive atoms are liable to decay. One decays now, the other may not decay for a further several thousand years. It is not possible to predict when a particular atom will decay. Nevertheless, the proportion of the atoms decaying in a given period is known precisely.

e^-
ν^-

The future is a mix of precisely determined events, consequences of random actions and some possible, as yet unknown, effects.

INDEX

Numbers shown refer to topics.

A

ABO system (blood), 53

Abrasion, 9

Abrasive, 41

Absolute zero, 1, 35

Absorption, 19, 40

AC (alternating current), 24, 26

Acceleration, 67

Accelerometer, 25

Acceptor impurity, 28

Acid (stomach), 52

Actin, 51

Additives (food), 54

Adenine, 56

Adenosine diphosphate (ADP), 45, 49

Adenosine triphosphate (ATP), 45, 48, 49, 51, 55, 59, 60

ADP (adenosine diphosphate), 45, 49

Adrenal Glands, 58

Adrenaline, 58

Aerial, 25

Africa, 8, 14, 16

Agriculture, 16

Air (inhalation), 52

Air conditioning system, 21

Air masses, 12

Air movement, 13, 21

Air pressure, 13

Aircraft, 34, 42

Algae, 14

Alien, 47

Alligator, 15

Alloy, 22, 23, 40, 42, 44

Alpha Centauri, 4

Alpha decay, 30

Alternating current (AC), 24, 26

Alumina, 41

Aluminium oxide, 41

Aluminium, 10

AM (amplitude modulation), 27

America, 16

Amino acid, 45, 46, 49, 54, 55, 57

Ammonite, 14, 15

Amorphous, 35, 38, 39

Amphibian, 14, 15

Amplification, 26, 27, 28

Amplitude modulation (AM), 27

Anaerobic bacteria, 48

Analogue information, 27

Analogue processing, 26

AND gate, 26

Androgens, 58

Andromeda, 3

Angiosperm, 15

Animal, 14, 15, 16, 30, 39, 44, 45, 48, 49, 50, 51, 53, 57

Animals, 50

Annelid, 15, 50

Antarctica, 8

Antenna, 18

Anthrax, 48

Anthropic principle, 70

Antibody, 52, 53, 55

Anticyclone, 12, 13

Anti-diuretic, 58

Antielectron, 1, 18

Antigen, 53

Antigravity, 2

Antilepton, 17

Antineutrino, 1, 17, 30

Antioxidant, 54

Antiparticle, 17, 70

Antiquark, 1, 17

Anti-wear parts, 42

Apatite, 41

Ape, 14, 15, 16

Apple, 54

Arachnid, 15

Aramid, 42

Arc lamp, 22

Archaean, 14

Aridity, 12

Arsenic, 28

Artery, 52

Arthropod, 15, 50

Asdic, 24

Asia, 8, 16

Asphalt, 36

Asteroid, 6, 7, 46, 47

Astrobiology, 47

Atmosphere, 6, 7, 9, 11, 12, 13, 14, 44, 47

Atom, 11, 17, 18, 19, 20, 21, 22, 23, 28, 32, 33, 35, 38, 39, 40, 41, 43, 44, 62, 65, 71

Atomic bomb, 31

Atomic force microscope, 43

Atomic Nuclei, 29

Atoms, 32

ATP (adenosine triphosphate), 45, 48, 49, 51, 55, 59, 60,

Audio tape, 25

Audio, 34

Auditory area (brain), 61

Auditory nerve, 60

Aurora, 11

Australia, 8, 14

Australopithecus Afarensis, 16
Australopithecus Africanus, 16
Autotroph, 45

B

Baboon, 16
Backbone, 15
Bacteria and Viruses, 48
Bacteria, 15, 44, 45, 48, 52, 57
Bakelite, 39
Balance (sense), 60, 61
Bamboo, 42
Bar code reader, 25
Barium, 29
Barometer, 25
Barrier island, 9
Baryon, 17
Basalt, 8, 10
Base pair (DNA), 56, 57
Bat (animal), 14, 34
Battery, 24
Beach, 9
Beaufort Scale, 13
Bell, 22, 25
Beta decay, 30
Big Bang, 1, 2, 3, 5, 18, 61, 70
Big Crunch, 2
Bile, 52, 55

Binary number, 27, 65
Binary star, 4, 67
Biological material, 33, 42
Biological processes, 11
Bird, 14, 15
Birefringence, 37
Biscuits, 54
Bit (information), 65
Black and white vision, 60
Black dwarf, 5
Black hole, 3, 4, 5, 67
Bladder, 52
Blob (string theory), 63
Blood cell, 53, 55
Blood clot, 53
Blood group, 53
Blood transfusion, 53
Blood, 50, 51, 52, 53, 58
Blue sky, 19
B-Lymphocyte, 53
Boat, 42
Body Circulation System, 52
Body colour (reflection), 38, 40
Boiling, 35
Bomb, 31
Bond (chemical), 33, 35, 36, 44
Bonded materials, 42

How Everything Came About

Bone marrow, 51, 53

Bone, 42, 50, 51, 58

Boron, 42

Boson, 17

Bottom quark, 17

Bottom-up technology, 43

Brachiopod, 15, 50

Brain, 16, 50, 59, 60, 61

Brass, 40

Brazil, 71

Breasts, 58

Breeze, 13

Brick, 41

Brittleness, 38, 41

Bronze, 40

Bubble formation, 35

Bulb (light), 22

Bulb (plant), 49

Burial of dead, 16

Butter, 54

Butterfly, 71

C

Cable (electric), 22

Calcite, 41

Calcium, 10, 51, 54, 58, 59

Calculator, 24, 27, 36, 37

Calorific value, 54

Cambrian Explosion, 14

Cambrian, 14

Camel, 14, 15

Camera, 19

Cancer cell, 53

Capacitor, 25, 26, 65

Capillaries (blood), 51, 52

Car battery, 24

Car, 42

Carbohydrate, 44, 53, 54, 55

Carbon dioxide, 10, 11, 14, 44, 45, 49, 50, 52, 55

Carbon fibre, 39

Carbon nanotube, 43

Carbon, 26, 29, 30, 31, 33, 39, 44, 45, 47, 54

Carbonation, 9

Carboniferous, 14

Carrier wave, 27

Cartilage, 51

Cat, 14

Catalyst, 24, 44, 52

Catalytic converter, 44

Cave, 16

Cell (biology), 30, 43, 45, 48, 49, 50, 51, 52, 53, 55, 56, 57, 58, 59, 60, 61

Cell (electric), 24

Cell respiration, 45

How Everything Came About

Cellophane, 39
Cellular immunity, 53
Celluloid, 39
Cellulose, 39, 49
Cemented tungsten carbide, 42
Centipede, 15
Central nervous system (CNS), 59
Ceramic, 41, 42
Cerebellum, 61
Cerebrum, 61
Cermet, 42
Chaos theory, 71
Charmed quark, 17
Cheddar cheese, 54
Chemical engineering, 44
Chemical industries, 44
Chemical plant, 21
Chemical Processes, 44
Chemical processes, 44, 62
Chemical production, 22
Chemical properties of atoms, 21, 32
Chemoreceptor, 60
Chicken, 54
Chickenpox, 48
Chimpanzee, 16
Chip (integrated circuit), 26
Chlorine, 22

Chlorophyll, 44, 49
Chloroplast, 45
Chocolate, 54
Cholera, 48
Cholesterol, 54, 55
Chromatid, 56
Chromosome, 15, 49, 56, 57
Circulation system), 50, 52
Cisterna, 51
CJD, 48
Clam, 15
Clay, 41, 44
Cleavage, 41
Cliff, 9
Climate, 12
Climatic regions, 12
Cloning, 49, 57
Cloud, 13
Cluster (galactic), 3
CNS (central nervous system), 59
Coal, 10
Coastal features, 9
Coastal influences, 13
Cobalt, 23, 42
Cochlea, 60
Code, 65, 66
Coelenterate, 15, 50

Coherent light, 20

Coil (electricity), 22, 24, 26

Cold (sense), 60, 61

Cold climatic region, 12

Cold front, 13

Collagen, 51

Collapse of wave function, 64, 65

Colloidal solution, 36

Colorant (food), 54

Colour force, 17

Colour vision, 60, 61

Colour, 18, 19, 40, 43,

Coloured glass, 38

Combustion, 44

Comet, 6, 14, 47

Communication system, 19, 20

Compact disc player, 25

Compact disc, 20

Composite materials, 41, 42, 43

Composite Materials, 42

Compound, 28, 33, 40, 41, 44, 54

Compression, 38, 40

Computer monitor, 21

Computer, 25, 27, 43, 61, 65

Concrete, 42

Condensation, 35

Conduction (heat), 35

Cone (eye), 60

Conifer, 15

Consciousness, 61

Continent, 14

Continental drift, 8, 14

Continental plate, 8

Convection cell (climate), 12

Convection, 8, 12, 35

Cooler, 24

Cooling fan, 25

Co-polymer, 39

Copper, 10, 29, 33, 38

Copper-nickel, 40

Coral, 15, 50

Core (Earth), 8, 14

Coriolis Effect, 12, 13

Cornflakes, 54

Corrosion, 22, 42, 44

Corticosteroids, 58

Corundum, 41

Cosmetics, 36

Cosmic background radiation, 1, 18, 70

Cotton, 39

Covalent bond, 28, 33, 39, 41

Cream (dairy), 54

Creativity (brain), 61

Credit card, 25
Cretaceous, 14
Crispbread, 54
Crisps, 54
Cristae, 45
Cro-Magnon, 16
Crust (Earth), 8, 10
Cryptography, 65, 66
Crystal glass, 38
Crystal, 10, 22, 23, 24, 28, 33, 39, 40, 41, 42
Crystalline material, 20, 23, 35, 41
Crystallization, 38, 39
Curie temperature, 23
Cutting (plant), 57
Cutting tool, 42
Cutting, 20
Cyclone, 12, 13
Cytoplasm, 45, 51, 55, 57
Cytosine, 56

D

Dark energy, 2
Dark matter, 2
Data storage, 43
dB (decibel), 34
DC (direct current), 24, 26
Decay (biology), 44

Decay (radioactive), 29, 30
Decibel (dB), 34
Deciduous tree, 14
Decoherence, 65
Decomposing, 48, 49
Deep-ocean volcanic rift, 46, 48
Deflation, 9
Delta, 9
Deoxyribonucleic acid (DNA), 15, 45, 46, 48, 50, 56, 57
Depression (weather), 13
Dermis, 51
Desert climate, 12
Desert, 9
Detergent, 36
Determinism, 71
Deuterium, 29, 31
Devitrification, 38
Devonian, 14
Diagnosis, 43
Diamagnetism, 23
Diamond, 41
Digestive system, 48, 50, 52, 54
Digital display, 25
Digital information, 27
Digital processing, 26
Dimensional stability, 42

How Everything Came About

Dimensions (of space), 63
Dinosaur, 14, 15
Diphtheria, 48
Dipole bond, 33
Dipole charge, 36
Direct current (DC), 24, 26
Disease, 48, 49, 53, 57
Dispersion (light), 19, 20
Displacement sensor, 43
Division (plants), 15
DNA (deoxyribonucleic acid), 15, 45, 46, 48, 50, 56, 57
Dog, 14
Domain (magnetic), 23
Dominant gene, 57
Donor impurity, 28
Doppler Effect, 34, 67
Down quark, 17
Drake Equation, 47
Droplet formation, 35, 36
Drug delivery, 43
Drugs, 49
Ductility, 40
Duodenum, 52

E

$E = mc^2$, 31, 68
Ear drum, 60

Ear, 60
Earth History, 14
Earth, 4, 5, 6, 7, 8, 11, 12, 14, 30, 44, 46, 47, 67, 68, 69
Earthquake, 30
Earthworm, 15
Ebonite, 39
Echinoderm, 15, 50
Eclipse, 7
Egg (food), 54
Egg cell, 56
Egg, 15
Einstein, 67, 68
Elasticity, 40
Elastomer, 39
Electric charge, 13, 17, 18, 21, 23, 26, 63, 64
Electric current, 21, 22, 26, 28
Electric Currents, 22
Electric field, 18, 21, 22, 27, 36, 43, 37
Electric force, 70
Electric Power, 24
Electrical impulse (nerve), 59, 60
Electrical properties, 42
Electrical resistance, 22
Electrical signal, 25
Electricity, 22, 24, 28, 40, 44

How Everything Came About

Electrification, 21

Electrode, 21, 24

Electrolysis, 22, 24

Electrolyte, 22, 24

Electromagnet, 22, 25

Electromagnetic force, 17, 18

Electromagnetic spectrum, 18

Electron beam, 21, 62

Electron microscope, 21, 62

Electron spin, 23, 32, 64, 65

Electron, 1, 5, 11, 13, 17, 18, 19, 20, 21, 22, 23, 24, 27, 28, 30, 32, 33, 35, 38, 40, 43, 49, 62, 64, 65, 66, 71

Electronic Components, 26

Electronic Processing, 27

Electronic Systems, 25

Electrons, 21

Electroplating, 22

Electrostatic induction, 21

Electrostatic precipitator, 21

Electrostriction, 25

Element, 5, 10, 28, 29, 30, 31, 32, 33, 40, 41, 44, 47, 54, 62, 70

Elementary particle, 17, 63

Elementary Particles, 17

Elephant, 15

Embryo, 56

Emery, 41

Emulsifier, 54

Emulsion, 36

Endocrine system, 50, 52, 58

Endogenic process, 9

Endothermic reaction, 44

Energy (heat), 35, 36

Energy (mass equivalence), 31, 68

Energy (metabolism), 54, 55

Energy (nuclear), 31

Energy (photon), 18, 19,

Energy (photosynthesis), 49

Energy (sound), 34

Energy (sources), 43

Energy from food, 54

Energy level (electron), 18, 19, 20, 32

Energy level (proton and neutron), 29, 30

Energy quanta, 62

Engine components, 42

Engraving (decorative), 16

Entanglement, 64, 65, 66

E-number, 54

Enzyme, 49, 52

Eocene, 14

Epidermis, 51

Equivalence principle, 67

Erosion, 9, 10

Escape velocity, 67
Etching, 43
ETI (extraterrestrial intelligence), 47
Europe, 8, 16
Evaporation, 12, 36, 51
Evolution of Humans, 16
Evolution, 15
Evolution, 15, 16, 46, 47, 50, 57
Exobiology, 47
Exogenic process, 9
Exothermic reaction, 44
Expansion (thermal), 35
Expansion (Universe), 70
Expansion of the Universe, 2
Explosion, 44
Extraterrestrial intelligence (ETI), 47
Extraterrestrial Life, 47
Extremophile, 46, 48
Extrinsic semiconductor, 28
Eye colour, 57
Eye, 60

F

Faeces, 52, 54
Fat, 51, 52, 54, 55
Fatigue (metal), 40
Fatty acid, 55
Fault (geological), 10

Feet, 16
Female, 56
Fermentation, 44, 48
Fern, 14, 15
Ferrel cell, 12
Ferrite, 42
Ferromagnetism, 23
Fibre (dietary), 54, 55
Fibre optics, 27
Fibre, 39, 42, 50, 51
Fibreglass, 42
Fibril, 51
Filament, 42
Fire, 16, 44
First quarter (Moon), 7
Fish (food), 54
Fish, 14, 15
Fishing rod, 42
Fission (nuclear), 31
Fission reactor, 31
Flake tools, 16
Flat Universe, 2
Flavour enhancer, 54
Flaw detection, 34
Flesh, 51
Floppy disc, 25
Flora, 14

Flow (liquid), 35, 36

Fluorescence, 19

Fluorescent lamp, 22

Fluorite, 41

FM (frequency modulation), 27

Fog, 13

Food chain, 44

Food, 36, 45, 49, 50, 52, 54, 55

Food, 54

Force, 67

Forest, 15

Fossil, 14, 16

Fracture, 9, 38, 40, 41

Free will, 61

Freezing, 35, 38

Frequency modulation (FM), 27

Frequency, 18, 26, 27, 34, 62, 63

Frog, 15

Front (weather), 13

Frost, 10

Fruit, 49

Fuel cell, 24, 43

Full Moon, 7

Fundamental (vibration), 34, 63

Fungus, 15, 45, 48, 49

Furnace lining, 41

Furnace, 22

Fusion (nuclear), 5, 31

Fusion, 20

Future, 2, 71

G

Galaxies, 3

Galaxy, 2, 3, 4, 47, 70

Gale, 13

Gall bladder, 52

Gallium, 28

Gametes, 49, 56

Gamma decay, 30

Gamma ray, 18

Gas, 10, 14, 20, 22, 34, 35, 44, 70

Geiger-Muller counter, 25

Gel, 36, 51

Gelatine, 36

Gem, 41

Gene analysis, 43

Gene, 15, 50, 56, 57

General Relativity, 67

Generator, 22, 24

Genetic changes, 14, 15

Genetic code, 46, 57

Genetic engineering, 47, 48, 57

Genetic fingerprints, 57

Genetics, 57

Genome, 57

Geological period, 14

Germ, 52, 53

German measles, 48

Gibbon, 16

Gills, 50

Glacier, 9, 10

Gland, 58, 59

Glandular fever, 48

Glass cloth, 38

Glass fibre, 19, 38

Glass filament, 19

Glass, 10, 19, 20, 21, 36, 38, 39, 41, 42

Glasses, 38

Global Positioning Satellites (GPS), 67, 68

Glucagon, 58

Glucose, 44, 45, 49, 55, 58

Gluon, 17

Glycerol, 55

Glycine, 46

Glycogen, 55

Glycolysis, 49, 55

Gold, 10, 29, 44

Gold-copper, 40

Gold-silver, 40

Golf ball, 62

Golf club, 42

Gorilla, 16

Graphene, 43

Grain structure (metal), 40

Gramophone pickup, 24, 25

Grana, 45

Granite, 8, 10, 41

Granulocyte, 53

Graphite, 42

Grassland, 14

Gravel, 42

Gravitational force, 1, 17

Gravitational wave, 1, 67

Graviton, 1, 17, 63, 67

Gravity (sense), 60

Gravity, 1, 2, 3, 5, 7, 8, 9, 10, 11, 62, 63, 67, 70

Greenhouse effect, 11, 47

Grinding wheel, 41

Growth (biology), 46, 49, 56, 58

Guanine, 56

Gymnosperm, 15

Gypsum, 41

H

Hadley cell, 12

Hadron, 17

Hail, 13

Hair colour, 57

Hair, 39, 51

Half-life, 30

Ham, 54

Hand, 16

Hardness, 40, 41, 42

Harmonic, 34

Headphones, 24, 25

Hearing aid, 25

Hearing, 60

Heart, 50, 52

Heat (sense), 61

Heat and Temperature, 35

Heat, 19, 22, 25, 30, 35, 40, 44, 50, 52, 54

Heater, 24

Heisenberg's uncertainty principle, 62

Helium, 1, 3, 5, 29, 30, 31, 32

Heredity, 46

Heterotroph, 45

Higgs boson, 17

Higgs field, 17

Highland climatic region, 12

HIV, 48

Hole (electrical conduction), 28

Holography, 20

Homo Erectus, 16

Homo Habilis, 16

Homo Sapiens Sapiens, 16

Homo Sapiens, 16

Hoofed animal, 14

Hormone, 50, 52, 55, 58

Horse, 14, 15

Horsetail, 15

Hospital, 21

Hubble's law, 2

Human genome project, 57

Human Metabolism, 55

Human, 15, 16, 50

Humidity, 13

Humoral immunity, 53

Hurricane, 13

Hydraulic power, 36

Hydrogen bomb, 31

Hydrogen, 1, 3, 5, 24, 29, 31, 32, 39, 43, 49, 54, 70

Hydrothermal vent, 46, 48

Hypothalamus, 58

I

IC (integrated circuit), 26, 43

Ice age, 9

Ice cream, 54

Ice, 9, 13, 35

Iceberg, 9, 48

Igneous rock, 10

Image production, 25
Immortality, 61
Immune system, 48, 53
Immunisation, 53
Immunity, 53
Implant, 43
Indicator light, 25
Inductor, 25, 26
Industrial processes, 30, 44
Industrial radiography, 18, 30
Infection, 53, 55
Inflation (Universe), 2
Influenza, 48
Information processing, 25, 27, 43
Information storage and retrieval 20, 43
Infrared, 11, 35
Infrasonic, 34
Ingestion, 53
Inhalation, 53
Insect, 14, 15
Insulation (electrical), 23, 26
Insulation (thermal), 38, 41
Insulator (electrical), 28
Insulin, 57, 58
Integrated circuit (IC), 26, 43
Interference pattern, 20

Intermediate neuron, 59
Internal body sensor, 43
Internal sensing, 60
Intestine, 50, 52, 55
Intrinsic semiconductor, 28
Invertebrate, 15, 50
Iodine, 29
Ion, 11, 13, 22, 32, 33, 36, 60
Ionic bond, 33, 41
Ionic compound, 36
Ionisation, 25
Ionosphere, 11
Iron (dietary), 54
Iron oxide, 38, 44
Iron, 5, 10, 23, 24, 29, 31
Iron-nickel, 40
Isobar, 13
Isostasy, 9
Isotope, 29, 30, 31, 62
Itching, 60

J

Jam (food), 54
Jelly fish, 15
Junction diode, 26
Junction transistor, 26
Jupiter, 4, 6, 47
Jurassic, 14

K

Kangaroo, 15

Keratin, 51

Kevlar, 42

Kidney, 52, 58

Krebs (tricarboxylic acid or TCA) cycle, 49, 55

Krypton, 29

L

Lactic acid, 48

Lactobacillus, 48

Laminated glass, 38

Lamp shell, 15

Land, 8, 13, 14, 46, 47

Landforms, 9

Language, 61

Laniakea supercluster, 3

Laptop, 36, 37

Large intestine, 52

Laser printer, 21

Laser, 18, 20, 43, 47, 66

Lasers, 20

Last quarter (Moon), 7

Latent heat, 35

Lava, 10, 38

LDR (light dependent resistor), 28

Lead oxide, 38

Lead sulphate, 24

Lead, 29, 30

Lean pork chop, 54

Learning, 61

LED (light emitting diode), 28

Leech, 15

Legionnaire's disease, 48

Lemur, 16

Length contraction (relativity), 68

Lens (eye), 60

Lens, 19, 20

Leprosy, 48

Lepton, 17

Levitated train, 22

Life, 14, 45, 46, 47, 70

Life, 45

Ligament, 51, 60

Light bulb, 22

Light dependent resistor (LDR), 25, 28

Light emitting diode (LED), 25, 28

Light year, 2, 3, 4, 64, 68, 69

Light, 19

Light, 2, 5, 19, 20, 25, 38, 40, 43, 44, 49, 67

Lighting, 43

Lightning conductor, 21

Lightning, 13, 21

Limestone, 10

Linen, 39

Lipid, 49, 54, 55

Liquid, 20, 22, 34, 35, 36, 38, 39, 41, 71

Liquid crystal, 25, 36, 37

Liquid crystal Display (LCD), 37

Liquid Crystals, 37

Liquids, 36

Lithium, 29, 41

Liver, 52, 55

Lizard, 15

Local group (galaxies), 3

Logic gate, 26, 27, 65

Logic, 61

Loop (string theory), 63

Lorentz equations, 68

Loudspeaker, 24, 25

Lubricant, 43

Lunar eclipse, 7

Lungs, 15, 50, 52

Lymph gland, 52

Lymph node, 53

Lymph, 52

Lymphatic system, 52, 53

M

Magallenic Clouds, 3

Magma, 10

Magnesium, 10, 41, 54

Magnet, 22, 23, 41

Magnetic field (Earth), 8, 11

Magnetic field, 18, 21, 22, 23, 24, 25, 31

Magnetic material, 26

Magnetic properties, 42

Magnetic resonance imaging (MRI), 23

Magnetism, 23

Magnetosphere, 11

Magnetostriction, 23, 25

Magnolia, 15

Main sequence star, 5, 6

Male, 56

Mammal, 14, 15, 16

Mantle (Earth), 8, 14

Manual control, 25

Marble, 10

Marine creatures, 14, 15

Mars, 4, 6

Marsupial, 14, 15

Mass increase (relativity), 68

Mass, 2, 5, 17, 18, 21, 27, 29, 31, 47, 62, 63, 67

Mathematics, 61

Measles, 48

Meat, 51
Mechanoreceptor, 60
Medical aids, 42
Medical applications, 43
Medical diagnosis, 30, 34
Medical implant, 42
Medical radiography, 18, 30
Medical therapy, 30, 34
Meiosis, 49, 56
Melting point, 38, 41
Melting, 35
Membrane (string theory), 63
Memory, 61
Meningitis, 48
Mercury vapour lamp, 22
Mercury, 4, 6
Meson, 17
Mesopause, 11
Mesosphere, 11
Messenger ribonucleic acid (mRNA), 57
Metabolic rate, 58
Metabolism, 44, 46, 48, 55
Metal extraction, 22
Metal, 21, 22, 28, 40, 41, 42, 43, 44
Metallic bond, 33, 40
Metals and Alloys, 40

Metamorphic rock, 10
Meteor, 6, 11, 14
Meteorite, 46
Meteorological factors, 12, 13
Mica, 41
Microbe, 14, 53
Micro-organism, 44, 53
Microphone, 24, 25
Microscope, 19
Microwave, 18, 25
Mid-ocean ridge, 8
Milk, 36, 58
Milky Way, 3, 4
Millipede, 15
Mind, 61
Mineral, 9, 10, 41, 49, 50, 52, 54, 55, 37
Miocene, 14
Mitochondrion, 45, 55, 59, 60
Mitosis, 49, 56
Mobile phone, 25, 37
Mohs hardness scale, 41
Molecular bond, 19
Molecular motor, 43
Molecular robot, 43
Molecule, 11, 18, 23, 33, 34, 35, 36, 38, 39, 41, 43, 44, 45, 46, 47, 52, 56, 57, 60, 62, 65, 71, 37

Molecules, 33

Mollusc, 15, 50

Monitor screen, 25, 36, 37

Monkey, 14, 15, 16

Monocyte, 53

Monomer, 39

Montreme, 15

Moon, 7, 47

Morality, 61

Moss, 14

Motor (molecular), 43

Motor area (brain), 61

Motor neuron, 59, 61

Motor, 22, 24

Mountain, 8, 9, 13, 14, 30

mRNA (messenger ribonucleic acid), 57

M-theory, 63

Mucus, 53, 60

Multi-cellular life, 14, 15

Multiphase alloy, 40

Multiplexing, 27

Mumps, 48

Muon, 17

Muscle, 42, 45, 50, 51, 55, 58, 59, 60, 61

Music, 34

Musical instrument, 34

Mutation, 48, 57

Myosin, 51

N

NADP (nicotinamide adenine dinucleotide phosphate), 49

NADPH, 49

Nail (finger), 51

NAND gate, 26

Nanometre, 43

Nanotechnology, 43

Natural rubber, 39

Neanderthal, 16

Neap tide, 7

Neon lamp, 22

Neptune, 4, 6

Nerve impulse, 59

Nerve, 51, 59

Nervous system, 50, 58, 59, 60

Neuron, 59, 60, 61

Neutrino, 1, 17, 30

Neutron star, 5

Neutron, 1, 5, 17, 23, 29, 30, 31, 62, 70

New Moon, 7

Nickel, 23

Nickel-iron, 8

Nicotinamide adenine dinucleotide phosphate (NADP), 49

Nitrogen, 11, 30, 44, 54, 55

NMR (nuclear magnetic resonance), 23

Nociceptor, 60

Crystalline Solids, 41

Non-metallic solid, 23, 41

Non-specific immunity, 53

NOR gate, 26

North America, 8

NOT gate, 26

Note (music), 34

Noxious, 60

n-p-n junction transistor, 28

n-type semiconductor, 28

Nuclear Energy, 31

Nuclear magnetic resonance (NMR), 23

Nuclear processes, 31

Nuclear reaction, 18, 31

Nuclear reactor, 31

Nucleic acid, 46

Nucleus (atom), 1, 17, 18, 21, 23, 29, 31, 32, 41

Nucleus (cell), 15, 45, 49, 56, 57

Nylon, 39

O

Observation (collapse of wave function), 71

Ocean bed, 10

Ocean current, 13

Ocean, 7, 8, 14, 44, 46, 47

Oceanic plate, 8

Octave (music), 34

Octopus, 15

Oestrogen, 58

Oil-in-water emulsion, 36

Oligocene, 14

Operational amplifier, 26

Optic nerve, 60

Optical communication, 43

Optical computing, 43

Optical display, 43

Optical fibre, 19, 20, 38, 66

Optical processing, 27

Optical tweezers, 43

OR gate, 26

Orang-utan, 16

Ordovician, 14

Ore, 10

Organ (biology), 50, 51, 57, 58

Organic electronics, 43

Organic, 10, 33, 44, 45, 46, 47, 54, 37

Organism, 10, 45

Origin of Earth, 8, 46

Origin of Life, 46

Orion arm, 4
Orthoclase, 41
Oscillator circuit, 27
Oscilloscope, 21
Ovaries, 58
Oven, 22
Overtone, 34, 63
Ovule, 49
Ovum, 56
Oxidation, 9, 44, 54, 55
Oxide, 10, 23, 38, 44
Oxygen, 10, 11, 14, 23, 24, 38, 44, 45, 49, 50, 52, 54, 55
Oxytocin, 58
Oyster, 15
Ozone, 11, 44

P

Pain, 61
Palaeocene, 14
Pancreas, 52, 57, 58
Pangaea, 8
Parallel processing, 61, 65
Paramagnetism, 23
Paranthropus Boisei, 16
Paranthropus Robustus, 16
Parasite, 48, 49
Parathyroid glands, 58
Parent, 56, 57
Particle (elementary), 17, 18, 21, 62, 63, 64, 65, 69, 70, 71
Passive immunity, 53
PCB (printed circuit board), 26
Peanuts, 54
Peat, 10
Pebble, 9
Peptide, 55
Percussion instrument, 34
Periodic table, 29, 30, 31, 32
Permanent magnet, 23
Permian, 14
Perseus arm, 4
PET (positron emission tomography), 30
Phase (Moon), 7
Phase (state of matter), 40
Phase (wave motion), 18, 20
Phloem, 49
Phosphate, 45, 49, 54, 56
Phosphorescence, 19, 25
Phosphorous, 54
Photochemistry, 19
Photocopier, 21
Photodiode, 25, 28
Photoelectricity, 19, 25

How Everything Came About

Photon, 1, 11, 17, 18, 19, 20, 24, 27, 28, 30, 31, 35, 40, 49, 60, 62, 66, 67

Photons, 18

Photoreceptors, 60

Photosensitive coating, 21

Photosynthesis, 19, 44, 49

Photovoltaic effect, 24

Phylum, 15

Piezoelectricity, 24, 25

Pig, 15

Pigment, 60

Pitch (music), 34

Pituitary gland, 58

Pixel, 37

Placental, 15

Planet, 4, 6, 7, 47, 63

Planetary nebula, 5

Plankton, 14

Plant, 14, 15, 16, 30, 39, 44, 45, 48, 49, 57,

Plants, 49

Plasma (blood), 52, 55

Plasma (state of matter), 1, 31, 35

Plastic (polymer), 30, 38, 39, 42

Plasticity, 38, 40, 41

Plasticizer, 39

Plate tectonics, 8, 9, 30, 47

Platinum, 44

Platypus, 15

Pleistocene, 14

Pliocene, 14

Pluto, 4, 6

p-n junction diode, 28

Polar bond, 33, 36

Polar cell (climate), 12

Polar climatic region, 12

Polar molecule, 33

Polar region, 11

Polarised light 19, 66, 37

Polaroid sun glasses, 19

Pole (magnetic), 23

Polio, 48

Polishing, 22

Pollen, 49

Pollutant, 21, 44

Polycrystalline, 40, 41

Polyethylene, 39

Polymer, 39, 42, 45, 46

Polymers, 39

Porcelain, 41

Pork chop, 54

Portland cement, 36, 42

Position sensor, 25

Positron emission tomography (PET), 30

166

Positron, 1, 18, 30

Possum, 15

Potassium aluminium silicate, 41

Potassium, 10, 29, 54, 59

Power station, 21

Precipitation (atmospheric), 12, 13

Pregnancy, 58

Preservative (food), 54

Pressure (fluid), 36

Pressure (sense), 60

Pressure sensor, 25

Pressure, 25

Primary bond, 33

Primate, 15, 16

Printed circuit board (PCB), 26

Prism, 19

Probability, 62, 64, 66, 70, 71

Progesterone, 58

Prokaryote, 15

Prolactin, 58

Proprioceptors, 60

Protein, 39, 44, 45, 46, 48, 51, 53, 54, 55, 56, 57, 60

Proterozoic, 14

Proton, 1, 5, 17, 23, 29, 30, 31, 32, 62, 70

p-type semiconductor, 28

Pyruvate, 55

Q

Quantisation, 32

Quantum computing, 43, 64, 65

Quantum Computing, 65

Quantum cryptography, 64, 65, 66

Quantum Cryptography, 66

Quantum dot, 43

Quantum effects, 43, 61

Quantum Entanglement, 64

Quantum mechanics, 33, 62, 63, 64, 70, 71

Quantum Mechanics, 62

Quantum teleportation, 64

Quark, 1, 17

Quartz, 10, 41

Quartzite, 10

Qubit, 65

R

Rabbit, 15

Rabies, 48

Radar, 18, 25

Radiation, 1, 30, 35, 70

Radio wave, 11, 18, 25, 27

Radio, 18, 25, 47

Radioactive dating, 30

Radioactivity, 30

Radioactivity, 8, 14, 18, 25, 29, 30, 31, 71

How Everything Came About

Radiography, 18
Rain, 13
Rainbow, 19
Raindrop, 19
Random events, 71
Rayon, 39
Reactor, 31
Receptor cells, 60
Receptor, 61
Recessive gene, 57
Red blood cell, 53, 55
Red giant, 5
Red sunset, 19
Reflection, 19, 38, 40
Refraction, 19, 43
Refractory, 41
Refrigeration, 35
Reinforced materials, 42
Relativity, 67, 68, 69
Relay, 22, 25
Reproduction, 45, 46, 49, 50, 56
Reproduction, 56
Reptile, 14, 15
Resin, 39, 42
Resistance, 28
Resistor, 25, 26
Respiration, 15, 44, 45, 50, 52

Retina, 60
Rhesus system, 53
Ribonucleic acid (RNA), 45, 46, 48
Ribosome, 45, 57
Ribulose bisphosphate, 49
River, 9, 10
RNA (ribonucleic acid), 45, 46, 48
Rock, 8, 9, 10, 14, 38, 41, 44
Rocket, 42, 67
Rocks, 10
Rod (eye), 60
Rose, 15
Rubber, 39
Ruby, 41
Runner (plant), 49
Rust, 44

S

Saccharide, 54, 55
Sagittarius arm, 4
Salamander, 15
Saliva, 53, 60
Salmonellosis, 48
Salt, 9, 22, 52, 54, 58
Sand dune, 9
Sand, 9, 42
Sandstone, 10
Sap, 45

Sapphire, 41
Satellite, 20, 43
Saturn, 4, 6
Sausage, 54
Scale (music), 34
Scarlet fever, 48
Scattering of light, 19, 20
Scent, 60
Schrodinger's wave equation, 62
Scree, 9
Sea anemone, 15
Sea creatures, 15
Sea cucumber, 15
Sea urchin, 15
Sea, 12, 13, 14
Seal (animal), 14
Search for extraterrestrial intelligence (SETI), 47
Season, 12, 13
Secondary bond, 33
Security tag, 43
Sedimentary rock, 10, 14
Sedna, 6
Seed, 14, 49
Seismic sensor, 25
Self-awareness, 61
Self-assembly, 43

Semicircular canals (ear), 60
Semiconductor devices, 24, 26, 28
Semiconductor, 28, 43
Semiconductors, 28
Sense organ, 59
Sense, 50, 60
Sensory area (brain), 61
Sensory neuron, 59
Series processing, 61
SETI (search for extraterrestrial intelligence), 47
Sex organ, 58
Sexual characteristic, 58
Shale, 10
Shape (sense), 61
Shark, 14
Sheet thickness gauge, 30
Shell (animal), 14, 50
Shell (electron), 23, 28, 32, 33, 40
Shingles, 48
Shock wave, 34
Shoot (plant), 49
Shooting star, 6
Shore platform, 9
Sial, 8
Siberia, 16
Sight, 60

Silica, 10, 38, 41
Silicate, 8, 10
Silicon dioxide, 10, 38, 41
Silicon, 10, 28, 38
Silk, 21, 39
Silurian, 14
Silver, 29
Sima, 8
Simultaneous events, 68
Sirius, 4
Skeleton, 50, 51
Skin, 15, 50, 51, 53, 60
Slate, 10
Slip plane, 38, 40, 41
Slug (animal), 15
Small intestine, 50, 52, 55
Snail, 15
Snake, 15
Snow, 13
Soap, 36
Sodium chloride, 22, 33, 54
Sodium vapour lamp, 22
Sodium, 10, 22, 29, 41, 59
Soil, 44
Solar cell, 24, 25, 43
Solar eclipse, 7
Solar Panel, 24

Solar radiation, 11, 46
Solar system, 6, 7
Solar wind, 11
Solder, 26, 40
Solid solution, 40
Solid, 34, 35, 36, 38, 39, 40, 41, 42
Solution (chemical), 9, 36, 41
Sonar, 24, 25, 34
Sound intensity, 34
Sound level, 34
Sound wave, 19, 24, 34
Sound, 25, 27, 34, 60
Sound, 34
South America, 8, 14
Space colony, 68, 69
Space ship, 67, 68, 69
Space vehicle, 24
Space warping, 67
Space, 2, 5, 20, 34, 46, 47, 62, 67
Spacetime, 5, 67
Spaghetti, 54
Special Relativity, 68
Special relativity, 68, 69
Species (life), 15
Spectacles, 19
Speech area (brain), 61

Speed (velocity) of light, 2, 18, 22, 27, 31, 47, 67, 68, 69

Sperm cell, 56

Spermatozoon, 56

Spherulite, 39

Spin (elementary particle), 17, 21, 29, 32, 64, 65, 66

Spinal cord, 59

Spiral galaxy, 4

Spit (coastal), 9

Sponge, 15, 50

Spore (bacteria), 48

Spore (plant), 15, 49

Sporting equipment, 42

Spring tide, 7

Squid, 15

Stabiliser (food), 54

Star fish, 15

Star, 3, 4, 5, 31, 35, 47

Starch, 15, 45, 49

Stars, 5

Steam, 35

Steel, 40

Steel-reinforced concrete, 42

Steppe climatic region, 12

Sterilisation, 30

Stiffness, 42

Stimulus (nerve), 59

Stomach acid, 53

Stomach, 52, 55

Storm, 13

Strain transducer, 23, 25

Strange quark, 17

Stratopause, 11

Stratosphere, 11

Strength, 40, 42, 43

String instrument, 34

String theory, 62, 63

String Theory, 63

Strong nuclear force, 17, 29

Sugar, 48, 54, 55, 56

Sulphur, 39, 54

Sulphuric acid, 24

Sun, 4, 5, 6, 7, 11, 12

Sunlight, 7, 44, 49

Supercluster (galaxies), 3

Superconductivity, 22

Supernova, 5Superposition, 62, 65

Supersonic, 34

Surface tension, 35, 36

Surgery, 20

Surveying, 20

Suspension (chemical), 36

Sweat gland, 51

Switch, 26, 65

Switching, 22, 27, 28

Synapse, 59

T

Table salt (sodium chloride), 22, 33, 54

Tachometer, 25

Talc, 41

Taste bud, 60

Taste, 60

Tau, 17

TCA (tricarboxylic acid or Krebs) cycle, 49, 55

Tears (eye), 53

Telephone, 25

Telescope, 19

Television, 18, 21, 25, 36, 47, 37

Temperate climatic region, 12

Temperature, 1, 10, 11, 12, 13, 22, 28, 31, 35, 42, 44, 46, 51, 71

Tendon, 51, 60

Tennis racket, 42

Tension, 38, 40

Testes, 58

Testosterone, 58

Tetanus, 48

Texas, 71

The Atmosphere, 11

The Brain, 61

The Earth, 8

The Endocrine System, 58

The First 410,000 Years, 1

The Human Frame, 51

The Immune System, 53

The Milky Way, 4

The Moon, 7

The Nervous System, 59

The Senses, 60

The Solar System, 6

The Twin 'Paradox', 69

Thermal cycling, 9

Thermistor, 25, 28

Thermoelectricity, 24

Thermometer, 25, 36

Thermoplastic, 39

Thermoreceptor, 60

Thermosetting, 39

Thermosphere, 11

Thermostat, 35

Thunder cloud, 21

Thunderstorm, 13

Thymine, 56

Thyroid gland, 58

Tide, 7

Time dilation, 18, 67, 68, 69

Tin, 10

T-Lymphocyte, 53
Toad, 1518
Toner, 21
Tongue, 60
Top quark, 17
Topaz, 41
Top-down technology, 43
Tornado, 13, 71
Total internal reflection, 19
Touch, 60, 61
Toughened glass, 38
Toughness, 42
Toxin, 48, 53
Tracing of elements, 30, 43
Trade wind, 12
Transducer, 25, 27
Transformer, 24
Transistor, 28, 43
Transmission of light, 19
Transmitter substances (nerves), 59
Transparent material, 19
Transplanted cell, 53
Tree shrew, 16
Tree, 14
Triangulum, 3
Triassic, 14
Tricarboxylic acid (TCA or Krebs) cycle, 49, 55
Trilobite, 14, 15
Triple star system, 4
tRNA (transfer ribonucleic acid), 57
Tropical climatic region, 12
Tropical storm, 13
Tropopause, 11
Troposphere, 11
Tuber, 49
Tuberculosis, 48
Tungsten, 29
Turbine components, 42
Turtle, 15
TV (television), 18, 21, 25, 47
Twin 'paradox', 69
Typhoid, 48
Typhoon, 13

U

Ultrasonic cleaning, 24
Ultrasonic, 34
Ultraviolet, 11, 46, 47
Uncertainty principle, 62
Uncertainty, 71
Universe, 1, 2, 3, 18, 47, 70, 71
Unstable nucleus, 29, 30
Up quark, 17

How Everything Came About

Uranium, 29, 31
Uranus, 4, 6
Urea, 52, 55
Urine, 52, 58
Uterus, 58

V

Vacancy (electron), 28
Vaccination, 53
Van der Waals bond, 33, 39
Vapour, 35, 36
Vegetation, 9, 10, 12
Vein, 52
Velocity (speed) of light, 2, 18, 22, 27, 31, 47, 67, 68, 69
Vendian, 14
Venus, 4, 6
Vertebrate, 14, 15, 50
Vibration sensor, 24
Vibration, 18, 25, 34, 35, 36
Video tape, 25
Vinegar, 54
Violin, 63
Virgo, 3
Virtual photon, 18
Virus, 43, 45, 46, 48, 52
Viscosity, 35, 36
Viscous, 39

Vision, 60
Visual area (brain), 61
Vitamin, 54, 55
Volcano, 7, 8, 9, 10, 14, 38, 44
Voltage, 22, 24, 26, 28
Vulcanizing, 39

W

Warm front, 13
Warmth (sense), 60
Watch (chronometer), 24, 36
Water vapour, 11, 13
Water, 10, 13, 24, 33, 35, 44, 45, 47, 49, 52, 54
Water-in-oil emulsion, 36
Wave (ocean), 9, 10
Wave (photon), 18
Wave (sound), 34
Wave equation (Schrodinger's), 62
Wave function, 64
Wave packet, 62
Wavelength, 11, 18, 19, 21, 34, 38, 62
Weak nuclear force, 17
Weaponry, 20
Wear resistance, 42, 43
Weather, 11, 12, 13, 71
Weather, 13
Weathering, 8, 9, 10, 44

174

Welding, 20

Westerlies, 12

Wetting, 36

Whale, 14

Whisker (crystal), 42

White blood cell, 53, 55

White dwarf, 5

White light, 19

Whooping cough, 48

Wind instrument, 34

Wind, 9, 10, 13

Wool, 39

Wound, 53

X

X-ray tomography, 23

X-ray, 18

Xylem, 49

Y

Yeast, 48

Yellow fever, 48

Z

Zygote, 56

Printed in Great Britain
by Amazon